M & E HANDBOOKS

M & E Handbooks are recommended reading for examination syllabuses all over the world. Because each Handbook covers its subject clearly and concisely books in the series form a vital part of many college, university, school and home study courses.

Handbooks contain detailed information stripped of unnecessary padding, making each title a comprehensive self-tuition course. They are amplified with numerous self-testing questions in the form of Progress Tests at the end of each chapter, each text-referenced for easy checking. Every Handbook closes with an appendix which advises on examination technique. For all these reasons, Handbooks are ideal for pre-examination revision.

The handy pocket-book size and competitive price make Handbooks the perfect choice for anyone who wants to grasp the essentials of a subject quickly and easily.

THE M & E HANDBOOK SERIES

Statistics

W. M. Harper

ACMA, AMBIM

Third Edition

MACDONALD AND EVANS

Macdonald & Evans Ltd.
Estover, Plymouth PL6 7PZ

First Published 1965
Reprinted 1966 (twice)
Reprinted 1967
Reprinted 1968
Reprinted (with amendments) 1969
Second edition 1971
Reprinted 1972
Reprinted 1973
Reprinted 1974
Reprinted 1975
Reprinted 1976
Third edition 1977
Reprinted 1979
Reprinted 1980
Reprinted 1981

ISBN: 0 7121 1955 8

Printed in Great Britain by
Richard Clay (The Chaucer Press), Ltd.,
Bungay, Suffolk

Preface to the Third Edition

I AM a great believer in the principle of leaving well alone the engine that is running well. Consequently, when I came to write the second edition of this book, I left as much untouched as I could since it seemed to be functioning quite adequately as it was. Writing the third edition I find myself in the same position, and since following this principle of leaving alone where possible proved successful in that previous instance, I have adopted it yet again. As a result the reader of the earlier editions will again meet much of the material presented in the same form as before.

Time, however, brings changes in examination demands. Inevitably examiners probe even deeper into the finer points of the theory and so additions have proved necessary. In consequence the chapter on the t distribution and the χ^2 distribution has been expanded to the point where each distribution warrants a chapter to itself. Also the subtle distinction between a one-tail and a two-tail test can no longer be ignored and today's student must, therefore, tackle the issue. The modern employment of hand-held calculators is also recognised by including their required entry sequences in the Appendix on statistical procedures.

Probability theory

By far the most important change is the addition of a new section on probability theory. Questions involving this theory have recently been nosing their way into examination papers on statistics and my earlier cavalier manner of ignoring the topic can no longer be sustained. Although probability theory underlies the techniques used in statistical sampling, I believe that it is possible to understand and apply these techniques at an elementary level with no more than a commonsense grasp of probability. I suggest, therefore, that the student who is not actually called on to learn probability theory for his examination can omit this section. As for the student not so fortunately placed I must recommend to him a very careful study of the two chapters on this topic. Indeed, a wider study is probably advisable since owing to space limitations I have been able to do no more than sketch the theory involved. There is, though, a rather fuller treatment in my book *Operational Research* in this series, should he wish to build on the matter covered in this book without having to re-orientate himself to the approach of a different author.

Inflation

The rapid rate of current inflation renders all price figures in a book out of date before the book is even published. I bow to the inevitability of this and have made little attempt to avoid the problem, contenting myself with merely pointing out that the application of the techniques involved is unaffected by the antiquity of any monetary data.

Examination questions

Where past examination questions given in the previous edition prove adequate I have kept them despite their age; there is no virtue in using new data merely for the sake of it. As has just been implied, a technique is independent of the data. This ensures that the selection of questions available to the lecturer for discussion is one that he has already evaluated. He is, then, able to choose tried and tested questions when aiming to consolidate the theory he has taught. This, I feel, is a useful feature of a book of this nature. The new material is, of course, represented by new questions.

Answers to most of the questions in the Progress Tests are given in Appendix VI. Where for reasons of space it has not been practical to supply any answer, the question has been marked with an asterisk (*).

I gratefully acknowledge permission to quote from the past examination papers of the following bodies:

Building Societies Institute (B.S.I.).
London Chamber of Commerce and Industry (L.C.C.I.).
Local Government Examinations Board (L.G.E.B.).
Union of Lancashire & Cheshire Institutes (U.L.C.I.).
Institute of Hospital Administrators (I.H.A.).
Royal Society of Arts (R.S.A.).
Institute of Cost and Management Accountants (I.C.M.A.).
Association of Certified Accountants (A.C.A.).
University of London (B.Sc.).

Institute of Administrative Accounting (I.A.A.).
Northern Counties Technical Examination Council (N.C.T.E.C.).
Institute of Transport (I.O.T.).
Institute of Marketing (Inst. M.).

September 1977 W. M. H.

Contents

PART THREE: FREQUENCY DISTRIBUTIONS

PART FOUR: CORRELATION

PART FIVE: PROBABILITY

PART SIX: SAMPLING THEORY

List of Illustrations

PART ONE

INTRODUCTION TO STATISTICS

CHAPTER I

Nature and Interpretation of Statistics

NATURE OF STATISTICS

1. What is statistics? Essentially statistics is a scientific approach to information presenting itself in numerical form which enables us to maximise our understanding of such information.

The figures which result from statistical analysis are also referred to as "statistics." But the subject of statistics is wider than this, and can be broadly divided into two parts:

(*a*) *Descriptive* statistics, dealing with methods of *describing a large mass of data* (*see* Parts Two and Three).

(*b*) *Analytical* statistics, dealing with methods that *enable a conclusion to be drawn* from the data (*see* Parts Four and Six).

2. Importance of statistics. Only recently has it been realised that society need not be run on the basis of hunches or trial and error. The development of statistics has shown that many aspects of progress depend on the correct analysis of numerical information—particularly in economics and business. Increasingly *figures* have become the basis of rational decisions rather than hunches and events are proving that these decisions based on figures give better results.

All this has led to an unprecedented demand for figures—a sort of "numbers explosion." But figures have to be understood and correctly handled, and this is the task of statistics. It is a task that will grow rapidly in importance—in business, in government and in science. It may even be one of the factors which determine the future rate of human progress.

3. Plan of this book. The book is arranged in logical sequence which can be summarised as follows:

Part One: Basic concepts of statistics.

Part Two: Collection and presentation of figures which form the basis of statistical analysis.

Part Three: Different ways of presenting and describing a collection of figures relating to a single characteristic.

Part Four: Techniques for determining the relationship between two series of figures.

Part Five: The theory of probability.

Part Six: The technique of drawing conclusions about a whole population from a relatively small sample—vital in modern statistics.

Part Seven: Index numbers and time series.

Finally there are appendixes which include a collection of past examination questions, and suggested answers to certain questions in the Progress Tests.

4. Variables. In statistics, numbers are used to measure characteristics, e.g. height, weight, time, money, the number of marriages or houses, the number of blue-eyed males in France, etc.

A characteristic that is being measured is called a "variable." Thus, if we are measuring the weights of children, then "weights of children" is the variable.

There is no restriction on the number of variables we must have in any statistical analysis, though of course there must be at least one. In this book most of our attention will be directed to single variable analyses, though two-variable problems (such as the relationship between advertising and sales) is the subject of the whole of Part Four.

INTERPRETATION OF STATISTICS

5. Danger of wrong interpretation. One of the first things that students of statistics must recognise is that figures can very easily be interpreted wrongly. Sayings such as "You can prove anything with figures" have gained widespread circulation because they embody the bitter experience of people who have found themselves misled by incorrect deductions drawn from basically correct figures.

Consequently many people tend to distrust statistics, and to regard statisticians as naive and incautious. In fact, statisticians are trained:

(*a*) *to be extremely careful in selecting information* on which to base their calculations; see "Biased sources," 6 below.

(*b*) *to make only such deductions as are strictly logical;* see "Invalid arguments," 7 below.

6. Biased sources. One of the chief dangers facing a statistician is that the sources of his information may be biased. A statistician must therefore always ask himself such questions as:

(*a*) *Who* says this?
(*b*) *Why* does he say it?
(*c*) *What does he stand to gain* from saying it?
(*d*) *How* does he *know*?
(*e*) Could he be *lying*? Or *guessing*?

EXAMPLES

(1) A politician might say, "Seventy-five per cent of the votes were cast for me. This means that most people in this constituency want me as their MP." But what if less than two-thirds of the electorate voted?

(2) "Nine out of ten TV stars genuinely believe our product the best." But did they choose their ten stars carefully?

(3) "Nobody has put forward a justifiable complaint about our products." Who decides what is justifiable?

7. Invalid arguments. A source of information may be completely unbiased, and yet faulty reasoning produces a completely invalid argument based upon it. Statisticians, therefore, are much concerned to ensure that their deductions are strictly logical.

EXAMPLES

(1) A politician might say, "My party has doubled its votes since the last election, and this proves that its support is greater than ever." But his party may have trebled the number of candidates it put up since the last election. Or the last election may have produced an abnormally low poll.

(2) "This penny has come down heads five times running. The chances are in favour of it coming down tails at the next toss." This is not so. The laws of probability give tails a 50/50 chance at every toss, no matter what sequence preceded it.

(3) "More people die in bed than anywhere else. Bed, then, is the most dangerous place to be."

Sometimes the fallacy in an argument is easier to detect if one asks oneself: is there any other possible explanation?

EXAMPLES

(1) "Hospital records show that the number of people being treated for this disease has doubled in the last twenty years. Therefore twice as many people suffer from the disease now." Possibly, but the figures may merely indicate that more people now take hospital treatment for the disease, or that methods of diagnosis have improved.

(2) "Last year 700 employees produced 150 000 ladders. This year 650 employees produced 160 000 ladders. This shows we have increased our productivity." Or alternatively we may have decreased our ladder sizes!

THE MATHEMATICAL LANGUAGE OF STATISTICS

8. Mathematical language. The main purpose of a language is to enable people to exchange ideas with the minimum effort and maximum clarity. Mathematics has its own language, with a vocabulary of its own (symbols), for the purpose of communicating mathematical ideas.

Mathematics is particularly suitable as a language, as it often involves a series of operations, and the order of the operations can be indicated by the lay-out of the expression. For example, one important measure is the *standard deviation*, and the procedure for calculating it could be stated in writing thus: "To find the standard deviation, divide the number of items in the group into the sum of the squares of the differences between the value of each item in the group and the mean of the group—and then find the square root of this figure." How much easier and quicker it is to set out this statement in mathematical language:

$$\sigma = \sqrt{\frac{\Sigma (x - \bar{x})^2}{n}}$$

Unfortunately the mere sight of mathematical expressions is often enough to make students shy away from what appears to them incomprehensible. But mathematical language is actually very easy to learn, and the few symbols explained below are enough to make the mathematical parts of this book fully intelligible.

9. The language of statistics. The student need not concern himself unduly if he cannot follow some of the explanations to the symbols below at first reading. When the work involving these symbols is covered later in the book their meaning will be much clearer.

Symbols representing numbers

x This is a collective symbol meaning *all* the individual values of a variable. Strictly speaking, it stands for x_1 and x_2 and x_3, etc., where x_1 is the first value in the group, x_2 the second, x_3 the third, etc.

y This is an alternative symbol to x. It is used where there are two sets of variables and x has already been used to indicate the first.

$\bar{x}$ (called "bar x"). A bar over a variable symbol indicates that it represents the *arithmetic mean* of the values of that variable.

f This stands for "frequency," i.e. the number of times a given value occurs in a collection of figures.

n This stands for the number of items (or pairs of items) in a collection of figures.

σ (called *sigma*). This stands for "standard deviation." A small suffix (e.g. σ_x; σ_y) indicates which variable is being referred to.

d This stands for "deviation," and is the difference between two values—the appropriate values depending on the statistical technique being used.

r This stands for "coefficient of correlation."

r' This stands for "coefficient of rank correlation."

p This stands for "proportion." } When used in sampling
q This equals $(1 - p)$. } theory.

p This stands for the probability of a success. } When used in
q This stands for the probability of a failure. } probability formulae.

z This stands for the deviation of a value from the mean measured in standard deviations.

Other symbols

$\simeq$ This means "approximately equals."

! This means that one must list in descending order all the numbers from and including the number that precedes the ! sign—and then multiply all these numbers together. ("!" is called "factorial.")

Σ (also called *sigma*; it is the capital letter version of σ). This means "the sum of" and simply indicates that the numbers following it should be added. For example, Σx means: "Add up all the values in the group relating to the x variable." If the group is 6, 8 and 15, then Σx is $6 + 8 + 15 = 29$.

Sometimes before adding we must perform a prior operation, such as multiplying or subtracting. If multiplying, the symbols to be multiplied are written together with the Σ sign in front, e.g. Σxy means *first* multiply each pair of x and y's and *then add* the products.

If subtracting, a bracket is put round the symbols to be subtracted and the Σ sign written *outside* the bracket, e.g. $\Sigma(x - \bar{x})$ means *first* subtract $\bar{x}$ from each x and then add.

Finally, note that Σx^2 means "Square each x figure first, then add," while $(\Sigma x)^2$ means "Add the x's first, then square the sum."

It is *very* important that operations are done in their correct order.

Suffixes

Small suffixes to symbols are used for *identification* in cases where the symbol may be used in more than one context, e.g. x_1, x_2, etc., as illustrated in the x symbol above, or σ_x which means the standard deviation of a collection of x's.

Index number symbols

In addition to the other symbols, index numbers use a few more of their own:

p Price of individual items.
p_0 Price of individual items in base year.
p_1, p_2, p_3, etc. Prices of individual item in subsequent years.*
q Quantity of individual items.
q_0 Quantity of individual items in base year.
q_1, q_2, etc. Quantities of individual items in subsequent years.*
w Weight.

Note that Σpq means that the price and quantity of each item in turn are first multiplied together and the products then added.

* If the current year only is being compared with base year, then the suffix 1 indicates the current year.

10. Example of interpretation. To demonstrate the interpretation of this language the following important formula will be analysed:

$$r = \frac{\Sigma xy - n\,\bar{x}\,\bar{y}}{n\,\sigma_x\,\sigma_y}$$

This means that r, the coefficient of correlation, is found by the following calculation:

(*a*) Multiply each x and y together and then add the products (Σxy).

(*b*) Next subtract from this figure the product of: the number of pairs of items; the arithmetic mean of the x variables; and the arithmetic mean of the y variables.

(*c*) Finally divide the answer by the product of: the number of pairs of items; the standard deviation of the x values; and the standard deviation of the y values.

PROGRESS TEST 1

(*Answers in Appendix VI*)

1. Consider the following statements critically:

(*a*) "Five years ago the average stay of patients in this hospital was 21 days—now it is only 16 days. This shows that we now cure our patients more quickly."

(*b*) "Most car accidents occur within 5 miles of the driver's home. Therefore long journeys are safer."

(*c*) "10 per cent of the drivers involved in car accidents had previously taken X. A parallel survey of drivers *not* involved in accidents show that only 1 per cent had taken X. This shows that taking X is a contributory cause of accidents."

CHAPTER II

Accuracy and Approximation

ACCURACY

1. Accuracy. Complete accuracy in statistics is often impossible owing to:

(*a*) *Inaccurate figures*. In statistics it is only rarely that completely accurate figures can be used. For example, if articles are weighed there is a limit to accuracy of the scales. Or when a count is made, there may be an element of doubt. Thus, however carefully a census is carried out there are always some people omitted for one reason or another who should have been included, and so the final figure is not completely accurate.

(*b*) *Incomplete data*. Complete accuracy is also impossible where calculations are made from data lacking all the necessary information. For example, in a later chapter we shall discuss the calculation of an average from data where individual values are not given, only the number of items which fall within a given range. Such an average figure must inevitably be only an approximation—albeit a very close approximation—to the true figure.

2. Spurious accuracy. It is not enough merely to appreciate that complete accuracy is usually impossible: it is also important that *no claim* should be made for such accuracy where it does not exist. Students may assert that they never do make such claims, but they should realise that every figure they write makes a statement regarding the accuracy of that figure.

For example, to write 4.286 means that an accuracy of up to three decimal places is being claimed. It *must* mean that, for if the accuracy is only to two decimal places then the 6 is a wild guess, in which case it is pointless to include it. But since it *is* included, a reader will assume it does have meaning and that the writer of the figure is therefore claiming an accuracy of three decimal places.

From all this it follows that a figure should include only those digits which are accurate—otherwise a greater accuracy is implied than the figure really possesses. When a figure implies an accuracy greater than it really has, such accuracy is termed *spurious accuracy*. Not only is spurious accuracy pointless, but it can also mislead—sometimes seriously.

For instance, if a manufacturer is told that the estimated cost of an article is £1.33 and he signs a contract to sell a quantity at £1.50 each, he may regret his action when he learns that the cost price was based on an estimate of £4.00 for making three—give or take £1.00.

APPROXIMATION

3. Rounding. If *some* figures in a survey will prove not to be accurate, there is little point in recording the other figures with absolute accuracy. Consequently, when such figures are collected they are frequently *rounded*. For instance, in a survey of petrol sold at petrol stations the recorded figures may be rounded to the complete gallon, to tens of gallons, or even to hundreds of gallons. This means making a decision as to *how* to round the actual detailed figures.

There are three methods of rounding:

(*a*) *Round up*—If figures are to be rounded by raising them to, say, the next then gallons, an actual figure of 185 would be recorded as 190.

(*b*) *Round down*—Conversely, the figure can be rounded by reducing it to the previous ten. In this case 185 would be recorded as 180.

(*c*) *Round to the nearest unit*—The figure can be rounded to the nearest ten. If a figure is exactly half way between, such as 185, the rule is "round so that the rounded figure is an even unit." Rounding 185 to the nearest ten means we must choose between 18 tens (180) or 19 tens (190). Since the rule says choose the even, 18 tens, i.e. 180, is recorded.

4. Biased and unbiased error. Although the inevitability of error is accepted in statistics, it is important to know if such error is biased or unbiased. *Biased error* results when the errors tend to be "all one way," so that any resulting figures will be known to be too big or too small. Thus, if in a survey all figures are rounded down, then a biased error results since all the figures (other than any exact figures) are recorded below their true value. So a total of such figures, for example, would be below the true total. Another form of biased error arises when items are counted. A non-existent item is very unlikely to be counted, whereas an existing item could well be missed—so that the total count would be biased towards under-stating the true number.

When any error is as likely to be one way as much as the other (and to much the same extent), we talk about an *unbiased error*. The error resulting from rounding to the nearest unit is, therefore, an unbiased error since any resulting figures are as likely to be over-stated as they are under-stated.

5. Significant figures. *Significant figures* are the digits that carry real information and are free of spurious accuracy. How many digits are in fact significant in any individual amount depends on the degree of rounding that may have occurred at some earlier stage in the calculations or when the data was collected. The concept is perhaps best grasped by means of examples:

Calculated figure	*Stated figure* (**bold** *digits*) *when accuracy is to:*		
	4 significant figures	*3 significant figures*	*2 significant figures*
613.82	**613.8**	**614**	**610**
0.002817	0.00**2817**	0.00**282**	0.00**28**
3 572 841	**3 573** 000	**3 57**0 000	**3 6**00 000
40 000	**40 000**	**40 000**	**40** 000

It should be noted that zeros which only indicate the *place value* of the significant figures (e.g. hundreds, thousands, hundredths, thousandths are not counted as significant digits.

6. Adding and subtracting with rounded numbers. When adding or subtracting with rounded numbers it is important to remember that the answer cannot be more accurate than the least accurate figure.

EXAMPLE

Add 225, 541, *and* 800, *where the* 800 *has been rounded to the nearest* 100.

225 + 541 + 800 = 1566. But since the least accurate figure is to the nearest 100, the answer must also be given only to the nearest 100, i.e. 1600. Any attempt to be more exact can

only result in spurious accuracy (indeed, even the 6 in 1600 is suspect, for since 800 is subject to be possible error of 50 either way, the true answer lies between 1516 and 1616; i.e. 1500 may easily be a more accurate rounded figure than 1600).

7. Multiplying and dividing with rounded numbers. When multiplying or dividing with rounded numbers, it is necessary to ensure that the answer does not contain more significant figures than the number of significant figures in any of the *rounded* figures used in the calculation.

EXAMPLES

(1) *Multiply* 1.62 *by* 3.2 (*both rounded*).
$1.62 \times 3.2 = 5.184$. But since 3.2 is only two significant figures the answer can only contain two significant figures, i.e. 5.2.

(2) *Multiply* 1.62 *by* 2, *where* 2 *is an exact* (i.e. *not a rounded*) *figure*.
$1.62 \times 2 = 3.24$. Since 1.62 is correct to three significant figures, the answer can be left as three figures (but note that the 4 is still suspect. The 1.62 could represent any figure between 1.615 and 1.625. These doubled are 3.23 and 3.25—either of which may be closer the true answer than our 3.24).

8. Maxima and minima. The above rules are easy to remember and while usually quite suitable in practice are not always free from possible slight error—as has been demonstrated. If an *error-free* figure is required then the following procedure should be followed:

(*a*) Re-write each rounded figure twice—first giving it the *maximum* value it could represent, and then giving it the *minimum*.

(*b*) Next calculate two answers: one using the maxima and the other the minima. The true figure should be stated as lying between these two answers.

EXAMPLE

Add 15.04, 21 *and* 10.3 (*all rounded numbers*).

Maxima	*Minima*
15.045	15.035
21.500	20.500
10.350	10.250
46.895	45.785

Therefore the true answer lies between 45.785 and 46.895.

This can be alternatively stated as 46.340 ± 0.555 (46.340 being the midpoint of 45.785 and 46.895).

9. Absolute and relative error. The actual amount of an error is termed the *absolute error*. However, we are often not so much concerned with the actual amount of an error as with the size of the error relative to the total figure. Thus in measuring the distance between London and Sydney, an error of a kilometre is insignificant, whereas such an error in the distance between Dover and Calais could be serious. Consequently, computing the percentage that the absolute error bears to the total figure gives a measure of the *relative* error.

$$\text{Therefore, } \textit{Relative error} = \frac{\textit{Absolute error}}{\textit{Estimated figure}} \times 100$$

EXAMPLE

Find the maximum relative error in the previous Example.

$$\textit{Relative error} = \frac{0.555}{46.34} \times 100 = 1.2\%$$

PROGRESS TEST 2

(*Answers in Appendix VI*)

1. Add: 280 tonnes; 500 tonnes; 641 tonnes; 800 tonnes; 900 tonnes.

2. 1200 people (to the nearest 100) were found to buy a sack of potatoes every quarter. Each sack weighed 112 kg (to the nearest kg). Calculate the total weight of potatoes bought in a year.

3. Add the following rounded figures and state the answer *as accurately as possible*, on the alternative assumptions that (*a*) rounding was to the nearest digit, (*b*) the figures were rounded *up*: 2.81; 4.373; 9.2; 5.005.

PART TWO
THE COMPILATION AND PRESENTATION OF STATISTICS

CHAPTER III
Collection of Data

Before any statistical work can be done at all, figures must be collected. The collection of figures is a very important aspect of statistics, since any mistakes, errors or bias which arise in collection will be reflected in conclusions subsequently based on such figures.

Always remember that *a conclusion can never be better than the original figures on which it is based.* Unless the original figures are collected properly, any subsequent analysis will be, at best, a waste of time and possibly even disastrous, since it may mislead, with serious consequences.

POPULATIONS AND SAMPLES

1. Populations. Before one starts collecting any data at all it is very important to know exactly what one is collecting data about! Newcomers to statistical method often fail to appreciate the importance of this—anxious to ascertain, say, local TV-viewing habits they make a rapid door-to-door survey and present their results as the TV-viewing habits of the people in their local area. However, in actual fact they did *not* collect data from the people in the area—they collected data from people in the area *who were in when they called.* If the survey had been made in the evening this factor would have obviously influenced the results. Most of the people who were usually out in the evenings would not be in and so their viewing would be unrecorded. Yet these people in the very nature of things would have different TV-viewing habits from the people whose viewing was recorded.

We call the group of people (or items) about which we want to obtain information the *population.* Clearly, the population must be defined very carefully. Thus, if we wish to investigate the colours of all the cars on English roads then our population will be all the cars on English roads. Note this is *not* the same as all English cars—many cars on English roads are foreign and some English cars are on roads abroad.

Defining the population may prove, in fact, very tricky. For instance, in an inquiry relating to the number of housewives who go out to work, how should "housewife" be defined? If it is taken to mean a wife with no occupation other than housekeeping, the survey would show that *no* housewives go out to work! If it means a wife who keeps house, then the question arises whether a newly-wed in a one-room flat with her husband away on business nearly all the time is really a "housewife." And should we include the wife who has a maid to do most of the housework? Different interpretations will lead to different results in the analysis.

Another problem sometimes associated with the population is that its full extent may not be known; for example, the number of people in England with unsuspected diabetes. Obviously it is not easy to collect figures for such a population.

2. Samples. If our population is relatively small and easily surveyed we may well examine every item in the population. However, in practice populations are usually too big, or items too inaccessible, to enable the whole population to be examined. One may have to be satisfied with examining only a part, or sample, of the total population. A *sample*, then, is a *group of items taken from the population for examination.*

3. Sample frame. A list of the entire population from which items can be selected to form a sample is called a *sample frame.* Without a sample frame random sampling **(12)** is impossible and selecting a sample becomes very much a hit-and-miss affair.

4. Costs, accuracy and samples. Although it may appear to the student better to survey the whole population than rely on a small sample, this is often not so. First of all it costs considerably more to examine the whole population than just a small sample and *these higher costs could easily exceed the value of the survey results.* Thus, cost very frequently rules out examining the whole population.

Moreover, it has often been found that taking only a sample results in improved accuracy. The reason for this is that a small sample can be given very careful attention and measurements made with a high degree of accuracy. Examining the whole population, on the other hand, is such a major task that often unskilled investigators have to be brought in and this, coupled with the monotony of examining large numbers of items, leads to so many errors that the overall cumulative error is greater than the error inherent in using sample results to draw conclusions about the whole population.

METHODS OF COLLECTION

5. Methods outlined. We have seen that figures relating to a chosen population can be obtained either from the whole population or from a sample. Whichever approach is decided upon, one or a combination of the following methods can be adopted:

(*a*) *Direct observation*—For example, counting for oneself the number of cars in a car park.

(*b*) *Interviewing*—Asking personally for the required information.

(*c*) *Abstraction from published statistics.*

(*d*) *Postal questionnaire*—Sending a questionnaire by post and requesting completion.

6. Direct observation. This is the best method, as it reduces the chance of incorrect data being recorded. Unfortunately it cannot always be used, generally on account of the cost. It would be uneconomical, for instance, to follow a housewife around for a month in order to find out how many times she vacuumed the lounge, quite apart from the practical difficulties. At other times this method cannot be used because the information cannot be directly observed, e.g. where people would spend their holidays if they had unlimited money.

7. Interviewing. A disadvantage of interviewing is that inaccurate or false data may be given to the interviewer. The reason may be (*a*) forgetfulness, (*b*) misunderstanding the question, or (*c*) a deliberate intent to mislead.

For example, a housewife who is asked how much milk she bought the previous week may (*a*) have forgotten, (*b*) include—or fail to include—any milk bought by her husband, or (*c*) overstate the amount because she has a number of children and feels she ought to have bought more than she did.

Another disadvantage of interviewing is that if a number of interviewers are employed they may not record the answers in the same way as the investigator himself would have done.

For example, to the question "Did you watch the XYZ TV programme last night—yes or no?" the interviewer may get the answer, "Well, part of it, then someone came and we switched off." He may well record this as a "no" answer, while the investigator may be taking it that such answers are being recorded as "yes."

Different standards like this can easily result in the wrong conclusions being drawn from the survey. One way of overcoming this disadvantage is to train the interviewers very carefully so that they record data in *exactly* the same form as the investigator himself would record it.

8. Abstraction from published statistics. Any data that an investigator collects himself are termed *primary data* and, because he knows the conditions under which they were collected, he is aware of any limitations they may contain.

Data taken from other people's figures, on the other hand, are termed *secondary data*. Users of secondary data cannot have as thorough an understanding of the background as the original investigator, and so may be unaware of such limitations.

Statistics compiled from secondary data are termed *secondary statistics*. Obviously the compilation of such statistics needs care, in view of the possibility of there being special features concerning the earlier statistics, or the population concerned, which are not known to the compiler. For example, a table relating to unemployment may cover only *registered* unemployment; if this fact were not indicated, a subsequent investigation using the figures in conjunction with *unregistered* unemployment figures could result in quite false conclusions.

For this reason anyone wishing to use published statistics should consider the purpose for which they were originally compiled. In many Government publications, the statistics *are* compiled in the knowledge that they will be used in the production of secondary statistics. Such statistics are carefully annotated and explained so that users will not be misled, and secondary statistics may be prepared from them with reasonable confidence. However, many others are published in connection with a specific inquiry and it may be very dangerous to use them as a base for the compilation of secondary statistics.

9. Postal questionnaire. This is the least satisfactory method, for the simple reason that only relatively few such questionnaires are ever returned. A return of 15 per cent is often considered a good response for certain types of survey, although reminder notices can usually improve the percentage. Moreover, the questionnaires which are returned are often of little value as a sample, since they are frequently biased in one direction or another.

For example, a questionnaire relating to washing-machine performance will be returned mainly by people with complaints, who are only too pleased at the chance to air them. Satisfied users will probably not bother to reply. A conclusion, based on the returned questionnaires, that washing-machine performance was on the whole bad would therefore be a false conclusion.

Bias of this sort is rarely as obvious as in the above example, and so cannot be allowed for in any analysis of returned questionnaires. Postal questionnaires therefore are not recommended unless one of the following conditions applies:

(*a*) Completion is a legal obligation (e.g. Government surveys).

(*b*) The non-responders are subsequently interviewed in order to obtain the required information.

(*c*) An appropriate sample of the non-responders is interviewed and the results indicate that failure to reply *is in no way connected with any bias.*

10. Design of a questionnaire. If a questionnaire is to be used—either as a postal questionnaire or as a basis for interviewing—the following points should be observed in its design:

(*a*) Questions should be *simple.*

(*b*) Questions should be *unambiguous.*

(*c*) The best kinds of question are those which allow a *pre-printed answer* to be ticked.

(*d*) The questionnaire should be as *short* as possible.

(*e*) Questions should be *neither irrelevant nor too personal.*

(*f*) *Leading questions should not be asked.* A "leading question" is one that suggests the answer, e.g. the question "Don't you agree that all sensible people use XYZ soap?" suggests the answer "yes."

(*g*) The questionnaire should be designed so that the *questions fall into a logical sequence.* This will enable the respondent to understand its purpose, and as a result the quality of his answers may be improved.

RANDOM SAMPLES

11. The problem of bias. Taking a sample is not simply a matter of taking the nearest items. If worthwhile conclusions relating to the whole population are to be made from the sample it is essential to ensure as far as possible that the sample is free from *bias*, i.e. allowing a particular influence to have more importance than it really warrants.

Assume, for example, that we wish to know what proportion of Europeans are fair-haired. If we took a sample wholly from Stockholm, our conclusion based on it would be wrong, because Swedes are a fair-haired nation. Our sample would thus allow "fair-hairedness" to have an importance greater than is warranted, i.e. it would be biased towards "fair-hairedness."

Unfortunately it is not sufficient merely to ensure there is no *known* bias in our sample. *Unsuspected* bias can equally invalidate our conclusions. So the question arises as to how one can select a sample that is free even from unknown bias.

12. What is a random sample? The possibility of taking a sample having unsuspected bias can be reduced by taking a random sample. A *random sample* is a sample *selected in such a way that every item in the population has an equal chance of being included.* This is the only method of sampling which we can be confident is free from bias.

There are various methods of obtaining a random sample but they all depend on the selection being wholly determined by chance.

One may imagine such a selection being made by writing the name (or number) of each item in the population on a slip of paper and then drawing from a hat, as in a lottery, the required number of slips to make up the sample. Although this gives the idea behind "random selection," in practice great care has to be taken that no bias can possibly arise—for instance, in this method adjacent slips could conceivably stick together. Probably the best method of selection is to number all the items and then allow a computer to throw out a series of random numbers which will identify the items to be used in the sample.

13. Random samples are not perfect samples. Finally, students should appreciate that a random sample is not necessarily a good cross-section of the population.

Drawing the names of Europeans out of a hat *could* result in a sample containing all Swedes—though it is highly unlikely. Thus a random sample too can be one-sided, and does not *guarantee* a sample free from bias. It simply guarantees that the *method of selection* is free from bias. This is a rather subtle difference, but an important one.

NON-RANDOM SAMPLING

14. Obstacles to using random samples. Occasions often arise when the selection of a pure random sample is not feasible. These occasions arise:

(*a*) when such a sample would entail much expensive travelling for the interviewers;

(*b*) when "hunting out" the people selected would be a long and uneconomic task;

(*c*) when there is no sample frame. One cannot select a random sample of fair-haired mothers, for example, since there is no sample frame for these mothers.

These obstacles are overcome by using *multi-stage*, *quota* and *cluster* sampling respectively.

15. Multi-stage sampling. In this technique the country is divided into a number of areas, and three or four areas are selected by random means. Each area selected is again subdivided and another small sample of areas selected at random. This process may be repeated until ultimately a number of quite small areas in different parts of the country has been selected. A random sample of the relevant people within each of these areas is then interviewed.

Although the technique does involve sending interviewers to different parts of the country, once an interviewer is in an appropriate area he can carry out all his interviews with virtually no further travelling. This brings the cost of the survey within reasonable bounds.

16. Quota sampling. To avoid the expense of having to "hunt out" specific people chosen by a random sample, interviewers are told to interview all the people they meet, up to a given number, which is called their *quota.* Such a quota is nearly always divided up into different types of people (e.g. working class, upper class, etc.) with sub-quotas for each type. The object is to gain the benefits of stratification (*see* **19**).

17. Cluster sampling. In this technique the country is divided into small areas, much as with the multi-stage method. The interviewers are sent to the areas with instructions to interview every person they can find who fits the definition given (e.g. fair-haired mothers).

This sort of sampling could be applied where surveys of (say) home workshops, or oak trees, were required. It should be carefully distinguished from multi-stage sampling, where the object is to cut down costs. Generally, cluster sampling is used when it is the *only* way a sample can be found.

SYSTEMATIC AND STRATIFIED SAMPLING

18. Systematic sampling. This is simply a short-cut method for obtaining a virtually random sample. If a 10 per cent sample (say) is required, then the sample can be selected by taking every tenth item in the sample frame, providing there is no regularity within the frame such that items ten spaces apart have some special quality. If by mischance there happened to be some pattern in the frame that coincided with the sampling interval, the sample would be biased: for instance, every tenth house in a street might have a bay window and therefore be slightly more expensive. Such bias is not common, however, and systematic sampling can often be safely used.

19. Stratified sampling. It is important to note that none of the techniques so far mentioned is better than a pure random sample. Stratified sampling is *better* than purely random methods and must therefore be distinguished from the others. In order to use it, one has to know what groups comprise the total population, and in what proportions.

For instance, in a survey relating to wages in a particular industry there will be male and female workers. As the proportion of each can easily be found, stratified sampling can be employed. The technique involves the following steps (assuming in this case that the relevant proportions are 3:7):

(*a*) Decide on the total sample size (say 1000).

(*b*) Divide this into sub-samples *with the same proportions as the groups in the population* (300 and 700).

(*c*) Select at random from within each group (strata) the appropriate sub-sample (300 male and 700 female workers).

(*d*) Add the sub-sample results together to obtain the figures for the overall sample.

The reason why stratified sampling is an improvement over a pure random sample is that it lessens the possibility of one-sidedness. As we have seen, a random sample of Europeans *could* be composed wholly of Swedes. But if it were arranged that the sample should contain different proportions of each nationality according to the size of the country's population, a one-sided sample would be impossible.

PROGRESS TEST 3

(*Answers in Appendix VI*)

1. What comments have you to make on the following statement made in *Mack's Mag*?

"We thought we would like to learn something of the physical characteristics of our readers so ten newsagents were selected and an observer stationed at each. This observer noted down physical facts about every person who bought a copy of *Mack's Mag* and the results of this random sample will be published in next week's issue."

2. What sampling methods would you use to obtain the following information?

(*a*) Ages of Australian-born persons resident in the U.K.
(*b*) Health details relating to U.K. borough councillors.
(*c*) Cinema takings.
(*d*) The views of the public on Sunday trading.

CHAPTER IV

Tables

It is a psychological fact that data presented higgledy-piggledy are far harder to understand than data presented in a clear and orderly manner. Consequently, the next step after the figures have been collected is to lay them out in an orderly way so that they are more readily comprehended. A very good form of layout is one of columns and rows. Such a layout is known as a *table.*

EXAMPLES

The following data are the result of an imaginary survey into the cinema-going and television-viewing habits of adult males in Great Watchet. The same information is presented in two ways: first in narrative form, i.e. written sentences, and then in tabular form. It is obvious which is the clearer and more concise.

(1) *Narrative form*

A survey of adult males in Great Watchet, taken in September 1977, by personal interview, showed that 122 of the 2049 single men under 30 attended the cinema less than once a month; 1046 attended one to four times a month, and 881 more than four times a month. Of the single men 30 and over the respective figures were 374, 202 and 23, a total of 599. As regards television viewing, 830 of the single men under 30 viewed less than 15 hours a week, the other 1219 viewing 15 hours or more. For single men 30 and over these figures were 358 and 241 respectively. As regards the married men, 1404 of the under-30s attended the cinema less than once a month, 289 attended one to four times a month and 112 more than four times a month. For those of 30 and over the figures were 1880, 115 and 10 respectively. TV viewing figures showed that 1162 married men under 30 viewed less than 15 hours a week and the remaining 643, 15 hours or more. Of the "30 and over" group, 484 viewed less than 15 hours and 1521 viewed 15 or more hours a week.

(2) *Tabular form*

TABLE I. SURVEY OF CINEMA ATTENDANCE AND TELEVISION VIEWING AMONG ADULT MALES IN GREAT WATCHET SEPTEMBER 1977

	Single		*Married*	
	Under 30	30 *and over*	*Under* 30	30 *and over*
Cinema attendance:				
Less than once a month	122	374	1404	1880
1–4 times a month	1046	202	289	115
More than 4 times a month	881	23	112	10
Total	2049	599	1805	2005
Television viewing:				
Less than 15 hours a week	830	358	1162	484
15 hours or more a week	1219	241	643	1521
Total	2049	599	1805	2005

Source: *personal interview*

PRINCIPLES OF TABLE CONSTRUCTION

1. Imagination and common sense are needed. The construction of a table is in many ways a work of art. It is not enough just to have columns and rows: a badly constructed table can be as confusing as a mass of data presented in narrative form. Yet, as with a work of art, it is difficult to lay down precise rules that will apply to all cases. For this reason the student should construct his tables as common sense guides him—and the sounder his common sense, the better will his tables be.

On the other hand, there are certain principles which must be observed in the construction of *all* tables. Though imagination can often improve a table, to ignore any of these principles will only result in a loss of clarity and impact which might be compared with mumbling instead of speaking clearly.

2. The basic principle. Of all the principles of table construction, there is one which is basic: *construct it so that the table achieves its object in the best manner possible.* This means the student must ask himself at the very beginning, "What is the purpose of this table?" Some of the possible reasons for which a table may be constructed are:

(*a*) To present the original figures in an orderly manner.

(*b*) To show a distinct pattern in the figures.

(*c*) To summarise the figures.

(*d*) To publish salient figures which other people may use in future statistical studies. (Many Government statistics are produced for this purpose, and the student is warned against using such tables as models if his own table is produced for a different purpose.)

Frequently table construction involves deciding which columns of figures should be adjacent to each other. For example, would Table I be improved if "Under 30" and "30 and over" had been the main column headings, with each broken down into "Single" and "Married" sub-headings? The answer depends, as always, on the purpose of the table. If the activities of *age-groups* are to be compared, it is best as it stands. But if a comparison between men of different *marital status* were required, the change would be an improvement.

Other kinds of decision must also be made. For instance, what totals should be shown? Should percentages be added? Should some figures be combined, or even eliminated? Invariably, the basis of all these decisions is the purpose to which the table will be put.

3. Other principles of table construction. The following additional principles should also be observed:

(*a*) *Aim at simplicity.* This is vitally important. A table with too much detail or which is too complex is much harder to understand, and so defeats its own object. *Remember:* it is better to show only a little and have that understood than to show all and have nothing understood.

(*b*) *The table must have a comprehensive, explanatory title.* If such a title would be too long, then a shorter one may be used together with a sub-title.

(*c*) *The source must be stated.* All figures come from somewhere and a statement of the source must be given, usually as a footnote.

(*d*) *Units must be clearly stated.* It is possible to reduce the number of figures by indicating in the title or in the headings the number of thousands, or other multiples of ten, each figure represents (e.g. writing "£000" in the title indicates that all figures in the Table are in thousands of pounds).

(*e*) *The headings to columns and rows should be unambiguous.* It is very important there should be no doubt about the meaning of a heading. If a lengthy heading would be necessary to remove ambiguity, then a short heading may be used with a symbol referring the reader to a footnote containing a more detailed explanation.

(*f*) *Double-counting should be avoided.* If a table shows "People in X: 100," and also "People in X and Y: 500" then the 100 people in X appear twice in the table. This is "double-counting" and, as it is apt to mislead, should normally be avoided. Cumulative figures may, of course, be shown but should be identified as such.

(*g*) *Totals should be shown where appropriate.* They are used in a table for one of the following purposes:

(*i*) To give the overall total of a main class.

(*ii*) To indicate that preceding figures are sub-divisions of the total.

(*iii*) To indicate that all items have been accounted for (e.g. if a survey of 3215 people is presented in a table, "Total 3215" indicates that all the people surveyed appear in the table, i.e. there are no gaps).

(*h*) *Percentages and ratios should be computed and shown if appropriate.* Frequently figures in tables become more meaningful if they are expressed as percentages, or (less often) as ratios. In constructing a table, therefore, it is important to decide whether or not it can be improved in this way. If it can, additional columns should be inserted in the table and the percentages (or ratios) computed and entered. Such percentages and ratios are sometimes called *derived statistics*.

4. Advantages of tabular layout. Comparison of narrative and tabular form as in the Example at the beginning of this chapter shows that tabular presentation has several distinct advantages, quite apart from being more readily intelligible:

(*a*) It enables required figures to be located more quickly.

(*b*) It enables comparisons between different classes to be made more easily.

(*c*) It reveals patterns within the figures which cannot be seen in the narrative form (e.g. in Table I cinema attendance is mainly confined to the younger men).

(*d*) It often takes up less room.

PROGRESS TEST 4

(*Answers in Appendix VI*)

1. In 1958, 1090 million passenger journeys were made on British Railways, yielding £137.6 million in receipts. Of the journeys, 351 million were at full fares, 426 million at reduced fares and an estimated 313 million by season ticket holders.

By 1961, journeys at full fares had fallen by 78 million, reduced fare journeys were up by 9 million and season ticket journeys up by 4 million.

During the same period, receipts from full fares went up from £74.0 million to £79.7 million, those from reduced fares went up from £46.1 million to £54.1 million and season ticket receipts increased by £5.9 million.

Draw up tables which will summarise the above information and will bring out the principal changes between the two years, and compute derived statistics where appropriate. (*I.O.T.*)

2. Criticise the following table:

Castings	*Weight of metal*	*Foundry hours*
Up to 4 kg	60	210
Up to 10 kg	100	640
All higher weights	110	800
Others	20	65
	——	——
	290	2000

CHAPTER V

Graphs

Tables, as we have seen, make data easier to understand. A further improvement in this respect can often be obtained by representing the data *visually*. Such an improvement stems from another psychological fact—that people, not being computers, are able to see spatial relationships much better than numerical relationships.

For example, in comparing sales of A with sales of B in the following table, what conclusions can you draw?

	1971	*1972*	*1973*	*1974*	*1975*	*1976*
Sales of A (units)	1121	1230	1339	1452	1568	1681
Sales of B (units)	492	541	602	644	691	738

It takes a certain amount of study to see that sales of A increase each year by more units than sales of B. But it is obvious at once when the same data are shown visually: *see* Fig. 1.

Such visual presentation can take many forms, but they can all be divided into two main groups, *graphs* and *diagrams*. The

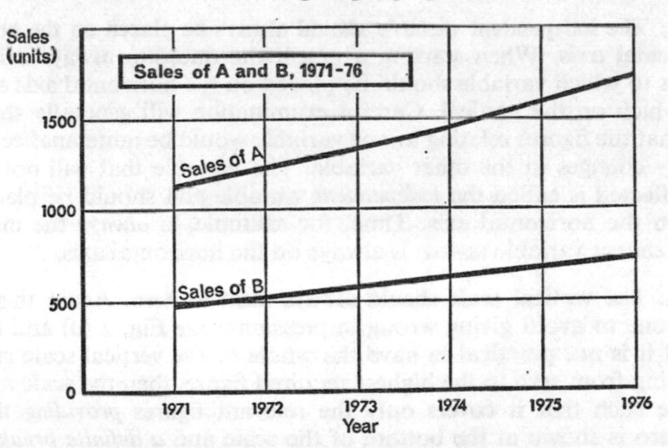

FIG. 1 *Graph as a means of comparing sales.*

difference between them is not very important; basically a graph is a representation of data by a continuous curve on squared paper, whilst a diagram is any other two-dimensional form of visual representation. Note that a line on a graph is always referred to as a *curve*—even though it may be straight. Diagrams are discussed in Chapter VIII.

PRINCIPLES OF GRAPH CONSTRUCTION

Graph construction, like table construction, is in many ways an art. However, like tables again, there are a number of basic principles to be observed if the graph is to be a good one. These are given below.

1. The correct impression must be given. Since graphs depend upon visual interpretation, they are open to every trick in the field of optical illusion. Note, for example, the difference between the impressions gained from the two graphs in Fig. 2 (*a*) and (*b*). They are one and the same graph; but in the second one the horizontal scale is only one-third the size of the first. Thus scale manipulation can considerably alter the dramatic impact of a graph. Needless to say, *good* (i.e. accurate, undistorted) presentation ensures that the correct impression is given.

2. The graph must have a clear and comprehensive title.

3. The independent variable should always be placed on the horizontal axis. When starting a graph the question always arises as to which variable should be placed on the horizontal axis and which on the vertical. Careful examination will generally show that the figures relating to one variable would be quite unaffected by changes in the other variable. The variable that will not be affected is called the *independent* variable and should be placed on the horizontal axis. Time, for example, is *always* the independent variable and so is always on the horizontal axis.

4. The vertical scale should always start at zero. Again this is done to avoid giving wrong impressions: *see* Fig. 2 (*a*) and (*c*). If it is not practical to have the whole of the vertical scale running from zero to the highest required figure, then the scale may be such that it covers only the relevant figures *providing* that zero is shown at the bottom of the scale and *a definite break in the scale is shown* (*see* Fig. 3).

5. A double vertical scale should be used where appropriate. If it is desired to show two curves which normally would lie very far apart, two scales may be put on the vertical axis, one curve being plotted against one scale and the other curve against the second scale. Figure 6, Chapter VII, is an example of a graph with a double vertical scale.

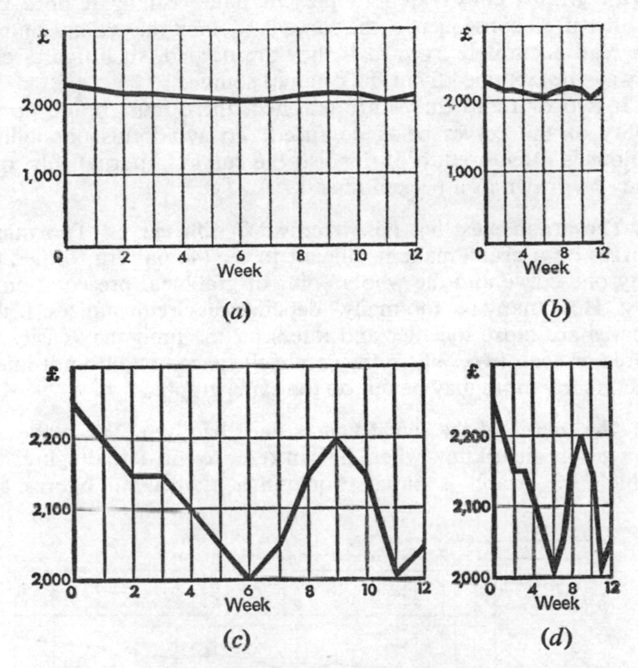

FIG. 2 *Effect of using different scales for the same graph.*

(*a*) Correctly scaled graph. (*b*) Graph with horizontal scale one-third of correct scale. (*c*) Zero omitted on the vertical scale. (*d*) Zero omitted and horizontal scale one-third of correct.

The student may not immediately appreciate that these four graphs involve identical figures. They should serve as a warning of the distortion that can result from badly chosen scales (*see also* Fig. 32).

6. Axes should be clearly labelled. Labels should clearly state both (*a*) the *variable*; and (*b*) the *units* (e.g. "Distance" and "Kilometres"; "Sales" and "£s"; "People viewing TV" and "Thousands").

7. Curve must be distinct. The purpose of a graph is to emphasise pattern or direction. This means that curves must be distinct. With graphs constructed to present data visually it does not matter if, in consequence, the curve is so thick that values cannot be read accurately from it. If they are needed, such figures can always be obtained from the original source.

If two or more curves are graphed, there must be no possibility of the curves being confused. To avoid this possibility, colour is often used to distinguish the curves. Alternatively, one may be drawn as a pecked line.

8. The graph must not be overcrowded with curves. Too many curves on a graph make it difficult to see the pattern formed by any one curve and the whole point of graphical presentation is lost. How many is "too many" depends on circumstances. If the curves are close together and intersect, the limit may easily be three or even two. Where they are well spaced and do not intersect, many more may be put on the same graph.

9. The source of the actual figures must be given. The reader of the graph must know where he can refer to the actual figures on which the graph is based. Sometimes the actual figures are

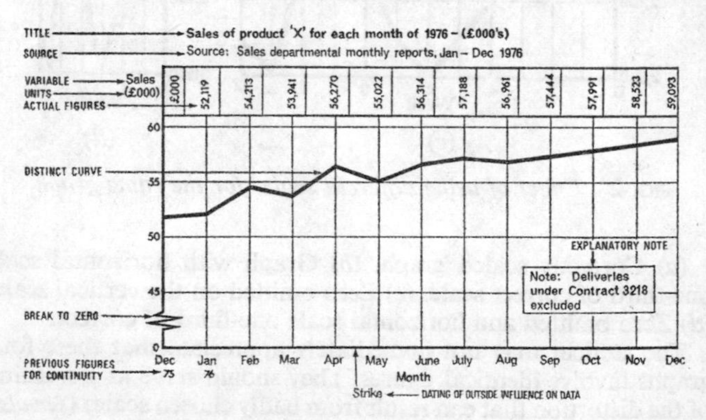

FIG. 3 *Model graph.*

inserted on the graph alongside the plotted points. Alternatively, they could be written at the top of the graph immediately over the point to which they relate (*see* Fig. 3).

TYPES OF GRAPHS

10. Presentation v. mathematical graphs. Students should note that two kinds of graphs can be constructed:

Presentation graphs. The purpose of these graphs is to present data in a visual form (e.g. the graphs discussed in this chapter).

Mathematical graphs. The purpose of these graphs is to enable estimates or predictions to be made from the graphs (e.g. regression line graphs and time series graphs discussed in Chapters XV and XXVI respectively).

The following slight, but important, differences arise in the construction of the two kinds of graphs and these the student should appreciate:

(*a*) *Curve thickness and type of graph.* If a graph is a *presentation graph* then it is important that the curve should be very distinct even though this means figures cannot be read accurately from it (*see* 7, where the point is further discussed). On the other hand if the graph is a *mathematical graph* it is very important that curves should be as thin as possible since the whole purpose of constructing such a graph is to enable previously unknown values to be read from the curves.

(*b*) *Plotting points on a time series graph.* A time series graph is one with time along the horizontal axis. When plotting values on a *presentation graph* of this sort the following rules should be observed:

(*i*) Plot *totals* at the *end* of the period to which they relate.

(*ii*) Plot *averages* at the *mid-point* of the period to which they relate.

However, when plotting time series values on a *mathematical graph* it is vitally important that one *always plots a value at the mid-point of the period to which it relates.*

If a student is uncertain as to which kind of graph he is dealing with, all he has to do is ask himself if the graph will be used to determine some otherwise *unknown* figure or not. If it will, the graph is a mathematical graph.

SEMI-LOG GRAPHS

11. When to use semi-log graphs. So far we have only considered the ordinary kind of graph, called the *normal scale graph.* It has one limitation in that it gives the wrong impression of rates of change. From the graph in Fig. 1, based on the figures given at the beginning of this chapter, it can be seen that sales of A have increased more than sales of B. But *the percentage increase for the 6 years is the same,* namely 50 per cent—sales of A in 1976 were 50 per cent greater than in 1971 and so were sales of B.

If then we want a graph that shows which product is increasing its sales at the faster rate we cannot use a normal scale graph. To get the right impression we need a *semi-log graph* (or *chart*) (*see* Fig. 4). This type has a *logarithmic scale on the vertical axis* (*see* Fig. 5). Note that the normal scale is retained on the horizontal axis—hence the term *semi*-log graph.

12. Features of a semi-log graph.

(*a*) If one curve is plotted on a semi-log graph (*see* Fig. 5):

(*i*) The slope of the curve indicates the *rate* at which the figures are increasing (decreasing).

(*ii*) If the curve is a straight line, *the rate of increase* (*decrease*) *remains constant.* Thus if the rate is 20 per cent per annum then the change will always be 20 per cent of the *previous* year's total (*see* curve *C* in Fig. 5).

Data — *Sales of A*

Year:	1971	1972	1973	1974	1975	1976
Sales:	1121	1230	1339	1452	1568	1681
Log:	3.0496	3.0899	3.1268	3.1620	3.1953	3.2256

Sales of B

Year:	1971	1972	1973	1974	1975	1976
Sales:	492	541	602	644	691	738
Log:	2.6920	2.7332	2.7796	2.8089	2.8395	2.8681

A good example of such a curve would be one showing compound interest. Although the amount of the interest is greater each year, it is always the *same percentage* of the total invested at the end of the previous period.

(*iii*) If the *absolute* increase is constant, the curve will become progressively less steep (and progressively more steep if an absolute *decrease* is involved). For instance, if the increase is always (say) £5000 per annum, the steepness must lessen, since

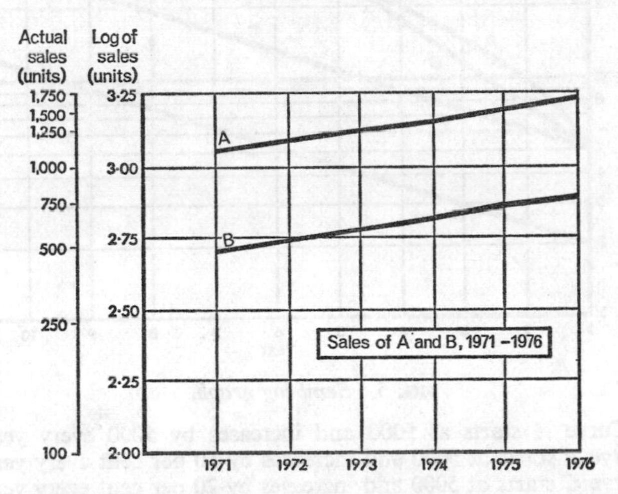

FIG. 4 *The same figures as in Fig.* 1 *but plotted on a semi-log graph.* The two curves are virtually parallel, indicating that the *rate* of increase is approximately the same for both, i.e. 8–10 per cent p.a.

£5000 becomes a *continually smaller percentage of the increasing total figure* (*see* curve *A* in Fig. 5).

(*b*) If *two* curves are plotted on a semi-log graph:

(*i*) The curve with the greatest slope has the greatest rate of increase or decrease (*see* curves *B* and *C* in Fig. 5).

(*ii*) If they are parallel the rates of increase (decrease) are identical (*see* Fig. 4 and curves *C* and *D* in Fig. 5).

(*iii*) If both curves over any part of the graph rise (fall) through the *same vertical distance* (say 1 cm), then both sets of figures have increased (decreased) by the *same percentage*.

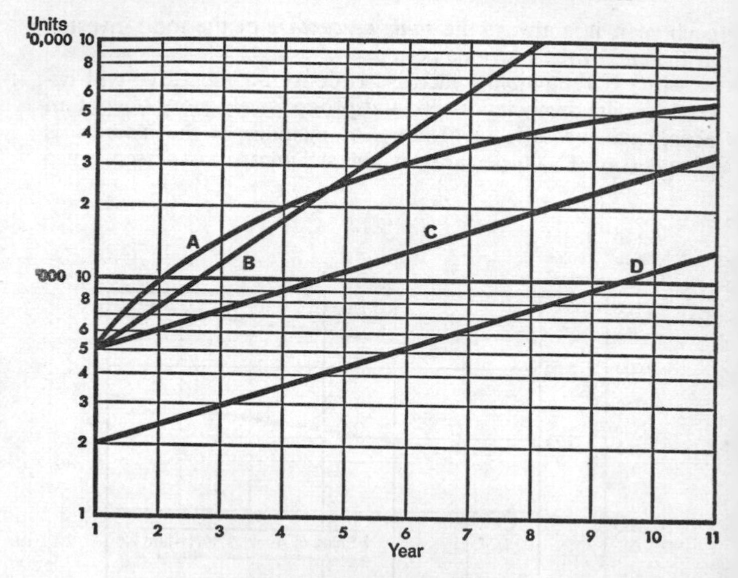

FIG. 5 *Semi-log graph.*

Curve *A* starts at 5000 and increases by 5000 every year.
Curve *B* starts at 5000 and increases by 50 per cent every year.
Curve *C* starts at 5000 and increases by 20 per cent every year.
Curve *D* starts at 2000 and increases by 20 per cent every year.
Note that although curve *D* is parallel to curve *C* the actual size of each year's increase is considerably smaller, e.g. in year 2, 400 as against 1000.

13. Construction of a semi-log graph. Semi-log graphs can be constructed in three ways:

(*a*) *By using semi-log graph paper* (which can be bought). Points are plotted in a straightforward way, although care is needed in reading the vertical scale (*see* Fig. 5).

(*b*) *By using a slide rule.* If the sliding section of a slide rule is removed, the printed scale can be used to mark out the *vertical* scale of a semi-log graph.

(*c*) *By plotting the logs of the variable.* If the log of each figure in the series is found from ordinary log tables and these log figures are plotted on a normal scale graph, semi-log curves are obtained (*see* Fig. 4). Note that actual values can also be shown on the vertical scale. This enables some idea to be obtained of the values of points on the curves.

NOTE: Students should remember that zero has no log and that therefore no attempt should be made to insert zero on the vertical scale of a semi-log graph.

14. Advantages of semi-log graphs.

(*a*) Semi-log graphs highlight rates of change.

(*b*) In addition, such charts allow a great range of values to be shown. Since the doubling of the vertical distance from the horizontal axis on such a graph is equivalent to *squaring* the value, it is possible to plot two widely separated series of figures—one series, say, around the 1000 level and another around the 1 000 000 level—on the same graph.

PROGRESS TEST 5

(*Answers in Appendix VI*)

1. The following (fictitious) figures relate to rainfall and the profits of an umbrella shop. Show both time series on the same graph:

Year	*1967*	*1968*	*1969*	*1970*	*1971*	*1972*	*1973*	*1974*	*1975*	*1976*
Rainfall (*cm*)	61	73	65	58	49	41	55	80	73	68
Shop profits (£)	6210	7400	8940	7730	7020	5910	4880	6610	9920	8630

2. Comment on the following graph:

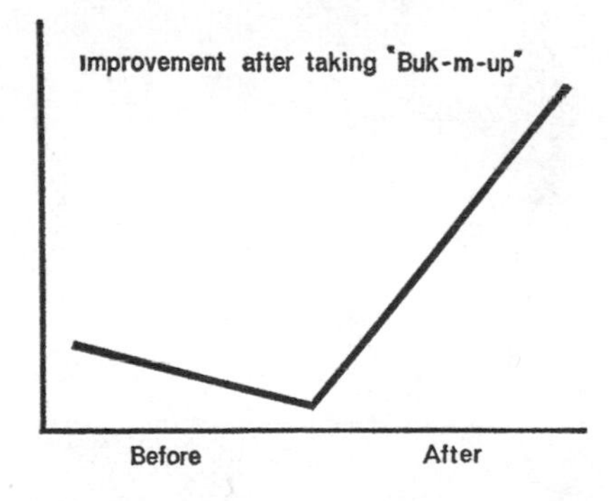

3. Graph the following data and then comment on the graph:*

Average temperatures at Little Fielding, 1919–76 (°c)

	Jan	*Feb*	*Mar*	*Apr*	*May*	*June*	*July*	*Aug*	*Sept*	*Oct*	*Nov*	*Dec*
Max.	8	9	11	13	16	18	18	17	15	12	10	9
Min.	−2	0	3	6	10	12	13	12	9	5	7	0

Source: *Little Fielding weather station records*

4. Plot the following time series on (*a*) a normal scale graph and (*b*) a semi-log graph (using, however, normal scale graph paper) and comment on the graphs:*

Year	*Group sales* (£000)	*Company A's sales* (£000)
1972	1620	135
1973	1780	154
1974	1950	176
1975	2140	195
1976	2350	208

CHAPTER VI

Moving Totals and Moving Averages

MOVING TOTALS

1. Insufficiency of period totals. Look at the following table. Do you think business is improving?

Monthly sales XYZ, 1976

Month	Jan	Feb	Mar	Apr	May	June
Sales (£)	4000	4100	4200	4300	4400	4500
Month	July	Aug	Sept	Oct	Nov	Dec
Sales (£)	4600	4700	4800	4900	5000	5100

On the face of it, business seems to be improving steadily. But, what if the sales for 1975 were as follows?

Month	Jan	Feb	Mar	Apr	May	June
Sales (£)	5000	5200	5400	5600	5800	6000
Month	July	Aug	Sept	Oct	Nov	Dec
Sales (£)	6200	6400	6600	6800	7000	7200

Clearly, business is *not* improving. Sales in January 1976 were £1000 below those of the previous January, and each month the gap between the sales for 1976 and the same month the year before increases, until by December the difference is £2100. So it can be seen that although each month in 1976 is better than the months before, it is worse—and progressively worse—than the same month in 1975.

2. Moving total. Obviously the figures given at first were misleading, and the question arises as to how this sort of wrong impression can be avoided. Direct comparison of the figures for one month with the same month the previous year is a possible solution, but this sort of comparison does not allow an overall trend to be easily observed. A better solution is the use of a *moving total* (or a *moving average—see* 5).

Examination of the XYZ's sales figures shows that the business is seasonal—indeed, it was for this reason that the 1976 figures on their own gave a false impression. This problem of seasonal influence frequently arises in statistics and *an excellent method of eliminating such influences is to add together twelve consecutive months.*

Such a total is inevitably free of any seasonal influence since all the seasons, busy and slack, are included in the total. Moreover, if we add the 12 months immediately preceding the end of *each* month in the table we shall obtain a series of totals, one for

TABLE II. M.A.T. OF XYZ'S SALES

Year	*Month*	*Sales* (£)	M.A.T. (£)	*Notes on calculation*
1975	January	5000	—	There can be no M.A.T. until 12 months' figures are available
	February	5200	—	
	March	5400	—	
	April	5600	—	
	May	5800	—	
	June	6000	—	
	July	6200	—	
	August	6400	—	
	September	6600	—	
	October	6800	—	
	November	7000	—	
	December	7200	73 200	Total of sales Jan–Dec 1975
1976	January	4000	72 200	73 200 − Jan 1975 + Jan 1976
	February	4100	71 100	72 200 − Feb 1975 + Feb 1976
	March	4200	69 900	71 100 − 5400 + 4200 etc.
	April	4300	68 600	
	May	4400	67 200	
	June	4500	65 700	
	July	4600	64 100	
	August	4700	62 400	
	September	4800	60 600	
	October	4900	58 700	Add Jan–Dec 1976 to cross-check the accuracy of this final figure
	November	5000	56 700	
	December	5100	54 600	

each month. Each total will be the total for the year immediately preceding the end of that month, and the series is called a *moving total* or—more specifically in this case, since the totals are yearly totals—a *moving annual total* (M.A.T. for short).

NOTE: Totals need not be yearly totals: they can relate to any period of time. There are, for instance, 5-year and 10-year moving totals.

3. Calculation of a moving total. Calculating a moving total is simply a matter of adding the appropriate group of periods immediately preceding the end of each individual period. However, the actual computing work can be reduced if it is appreciated that, once the first total has been found, the next total will be the same except for the difference between the new period which has been added and the old period which has been dropped. Table II demonstrates this method of calculating the moving annual total for XYZ's sales.

4. Significance of a moving total. If a moving total for sales drops, it means the position is deteriorating, since such a drop indicates that the current period sales fail to equal sales for the same period the previous year. A continuing fall indicates a continuing failure of current sales to equal the previous year's sales. Conversely, a rising total indicates an improvement. Of course, if costs are being considered, then a declining total indicates improvement in the form of reduced costs.

It can be seen, therefore, that the use of moving totals helps to eliminate incorrect impressions. If such totals are graphed, the slope of the line gives a good indication of the immediate trend.

MOVING AVERAGES

5. Moving average. This is simply a moving total divided by the number of periods comprising that total.

For example, look at Table II again. It shows the M.A.T. throughout 1976. Since each total is the sum of 12 months, the *moving average* is each moving total figure divided by 12, i.e.

Month	M.A.T.	*Moving average*
December (1975)	73 200 ÷ 12 =	6100
January (1976)	72 200 ÷ 12 =	6017
February	71 100 ÷ 12 =	5925 etc.

6. Graphing moving averages. When graphing moving averages, care needs to be taken as to where the moving average points are plotted. Being *averages*, they must be plotted *at the mid-point of the period* of which they are the average. Thus the £6100 just calculated above must be plotted at 30th June 1975 and the £6017 at 31st July 1975. Note that moving totals, being *totals*, may be plotted at the *end* of the relevant periods (*see* V, **10**).

In order to assist the correct plotting of moving averages it is a good idea to write the average opposite the mid-point of its period when constructing the table of moving averages. For instance, the January 1976 figure of £6017 (from **5**) will be written opposite the half-way point between July and August 1975, thus:

Month	*Actual*	*Moving average*
July	6200	
		6017
August	6400	

7. Advantages of moving averages.

(*a*) They eliminate seasonal variations.

(*b*) When period figures fluctuate violently, moving averages smooth out the fluctuations.

(*c*) They can be plotted on the same graph as the period figures without a change of scale (a M.A.T. is usually so many times bigger than period figures that it is often impossible to put both meaningfully on a graph with only one vertical scale).

PROGRESS TEST 6

(*Answer in Appendix VI*)

1. From the following annual figures calculate (*a*) the 3-year moving total, (*b*) the 3-year moving average, (*c*) the 10-year moving average.

Plot these, together with the individual annual figures, on the same graph. What is the difference between the 3-year moving average curve and the 10-year moving average curve?

Yearly figures 1950–74 (*units*)

1950	5	1955	8	1960	20	1965	9	1970	18
1951	8	1956	15	1961	16	1966	15	1971	22
1952	6	1957	10	1962	15	1967	8	1972	16
1953	12	1958	10	1963	6	1968	12	1973	14
1954	4	1959	13	1964	18	1969	14	1974	20

CHAPTER VII

Z Charts and Lorenz Curves

Although graphs are generally designed for the particular purpose for which they are required, there are two types of graph so common that their forms have become standardised. They are Z charts and Lorenz curves.

Z CHARTS

1. Description of Z charts. A Z chart is simply a graph that extends over a single year and incorporates:

(*a*) Individual monthly figures.
(*b*) Monthly cumulative figures for the year.
(*c*) A moving annual total.

It takes its name from the fact that the three curves together tend to look like the letter Z.

2. Points in the construction of a Z chart.

(*a*) Very often a double scale is used on the vertical axis, one for the monthly figures and the other for the M.A.T. and cumulative figures. This is because the M.A.T. is some twelve times larger than the normal monthly figure. If the same scale were used for both curves it would mean that the curve of the monthly figures would tend to creep insignificantly along the bottom of the graph.

(*b*) As an example, Fig. 6 shows a Z chart for the XYZ's sales figures discussed in VI, **1**. Note where the different curves start:

(*i*) *Monthly figures* at the December figure of the previous year.

(*ii*) *Cumulative figures* at zero.

(*iii*) *M.A.T.* at the M.A.T. figure for the December of the previous year.

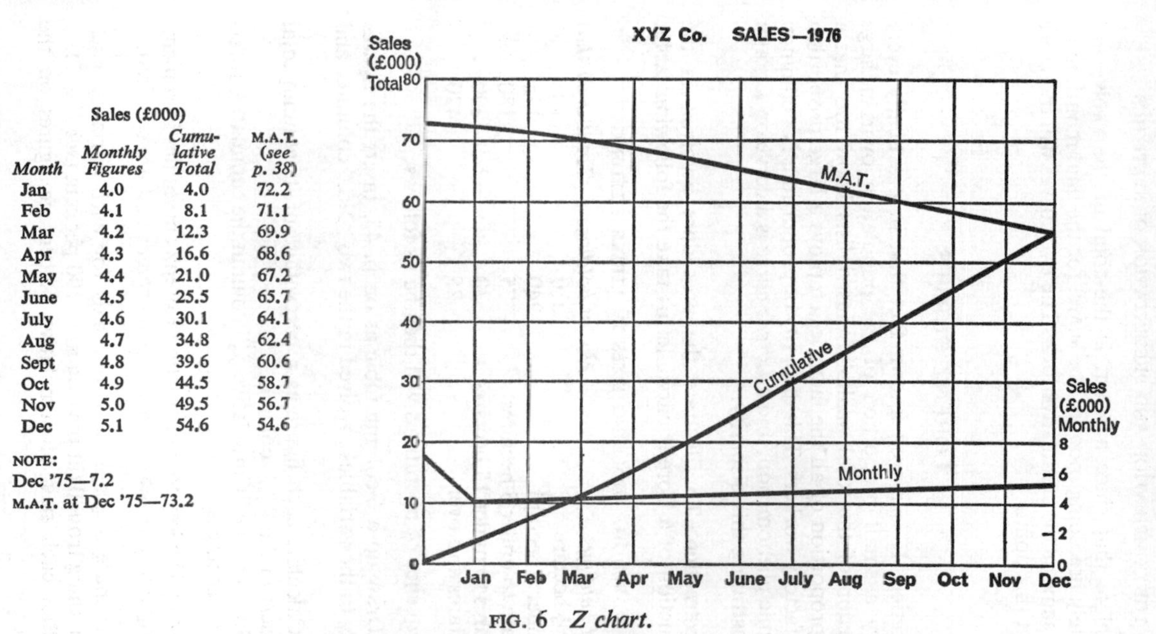

Sales (£000)

Month	*Monthly Figures*	*Cumulative Total*	M.A.T. (*see p. 38*)
Jan	4.0	4.0	72.2
Feb	4.1	8.1	71.1
Mar	4.2	12.3	69.9
Apr	4.3	16.6	68.6
May	4.4	21.0	67.2
June	4.5	25.5	65.7
July	4.6	30.1	64.1
Aug	4.7	34.8	62.4
Sept	4.8	39.6	60.6
Oct	4.9	44.5	58.7
Nov	5.0	49.5	56.7
Dec	5.1	54.6	54.6

NOTE:
Dec '75—7.2
M.A.T. at Dec '75—73.2

FIG. 6 *Z chart.*

NOTE (*i*) the double vertical scales, one for the M.A.T. and cumulative, the other for monthly, and (*ii*) that these figures are not typical; *see* Progress Test 7, Question 1, for more typical figures.

(*c*) Note that since a M.A.T. is the total of the twelve immediately preceding months, the M.A.T. for the final month must be the same as the cumulative total. The two curves will therefore meet at this point.

LORENZ CURVES

3. Function. It is a well known fact that in practically every country a small proportion of the population owns a large proportion of the total wealth. Industrialists know too that a small proportion of all the factories employs a large proportion of the factory workers. This disparity of proportions is a common economic phenomenon, and a *Lorenz curve* is a curve on a graph demonstrating this disparity.

4. Construction. To illustrate the procedure involved in the construction of a Lorenz curve, let us take the following table:

TABLE III. HOLDING SIZES IN LITTLE FIELDING

Size of holding	*No. of holdings*	*Total area (ha)*
Under ½ hectare	310	105
½ to under 1 hectare	240	175
1 hectare to under 5 hectares	75	180
5 hectares to under 15 hectares	30	300
15 hectares and over	25	420

The sequence of operations will then be as follows:

(*a*) Draw up a 6-column table as on p. 45. Insert the figures relating to the variables involved in the two "No." columns, and add.

(*b*) Calculate each figure as a percentage of its column total and insert in the "%" column.

(*c*) Insert in the "Cumulative %" column the cumulative totals of the percentages.

NOTE: The key to constructing Lorenz curves lies in remembering that it is the *cumulative percentages* that are required.

(*d*) Prepare a graph with one axis for each variable, each scale running from 0 (at the origin) to 100 per cent (*see* Fig. 7).

(*e*) Plot each pair of cumulative percentage figures on the graph.

(*f*) Starting at the origin, join these points in a smooth curve.

(*g*) Insert the "line of equal distribution" by joining the origin to the "100 per cent/100 per cent" point.

Holdings			Total area (ha)		
No.	%	*Cumulative* %	*No.*	%	*Cumulative* %
310	45.5	45.5	105	9	9
240	35.5	81	175	15	24
75	11	92	180	15	39
30	4.5	96.5	300	25.5	64.5
25	3.5	100	420	35.5	100
680	100		1 180	100	

5. The line of equal distribution. If, in our example, all the holdings had been of equal size, then clearly the total acreage of (say) 25 per cent of the holdings would be 25 per cent of the total acreage of holdings in Little Fielding. Similarly, 50 per cent of the holdings would constitute 50 per cent of the area, and 75 per cent of the holdings 75 per cent of the area. If these pairs are plotted on the graph they will be found to fall on a straight line running from the origin to the 100 per cent; 100 per cent point. Such a line, then, is *the curve which would be obtained if all holdings were of equal size.* It is called, therefore, the *line of equal distribution.*

6. Interpretation of Lorenz curves. The extent to which a Lorenz curve deviates from the "line of equal distribution" indicates the degree of inequality. The further the curve swings away, the greater the inequality. There is no actual measure of this inequality but its extent can be indicated by reading the curve at the point where it lies furthest from the line of equal distribution.

For example, in Fig. 7 the curve at its furthest point from the line of equal distribution is approximately at the 87 per cent "Holdings" and 30 per cent "Total area." This means that 87 per cent of the holdings contain 30 per cent of the total area—or, put the other way round, a mere 13 per cent of the holdings enclose 70 per cent of the total area.

7. Use of Lorenz curves. Lorenz curves can be used to show inequalities in connection with matters such as:

(*a*) Incomes in the population.
(*b*) Tax payments of individuals in the population.
(*c*) Industrial efficiencies.
(*d*) Industrial outputs.
(*e*) Examination marks.
(*f*) Customers and sales.

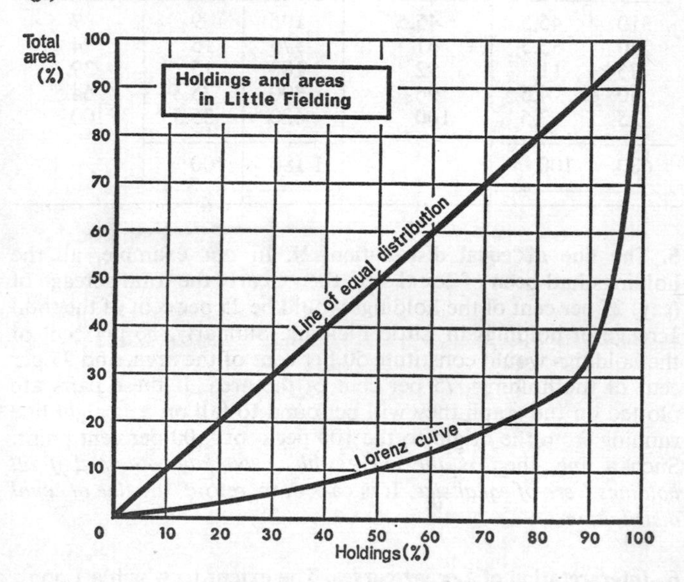

FIG. 7 *Lorenz curve.*

In some instances, Lorenz curves can be used to compare *two series* of inequalities. For instance, if a second Lorenz curve relating to holdings in, say, Great Fielding, were superimposed on the curve for Little Fielding in Fig. 7, then it would be possible to compare inequalities of land holding in these two communities and to see in which one holdings were nearer to being equally distributed.

PROGRESS TEST 7

(Answers in Appendix VI)

1. From the following data taken from monthly sales statistics construct a Z chart for 1976 and comment on the graph:

ALPHA LTD. SALES (£)

Month	1975	1976	Month	1975	1976
Jan	15 000	17 000	*July*	6 000	11 000
Feb	14 000	19 000	*Aug*	6 000	1 000
Mar	11 000	18 000	*Sept*	8 000	5 000
Apr	10 000	18 000	*Oct*	10 000	5 000
May	8 000	18 000	*Nov*	10 000	8 000
June	7 000	12 000	*Dec*	13 000	10 000

2. The following figures come from the Report on the Census of Production for 1958:

TEXTILE MACHINERY AND ACCESSORIES

Establishments	Net output
No.	£000
48	1 406
42	2 263
38	3 699
21	2 836
26	3 152
16	5 032
23	20 385
214	38 773

Analyse this table by means of a Lorenz curve and explain what this curve shows.

(I.C.M.A.)

CHAPTER VIII

Diagrams

Graphs are not the only way of presenting data visually. The most commonly used alternatives can be classified as follows:

(*a*) *Pictorial representations.*
- (*i*) Pictograms.
- (*ii*) Statistical maps.

(*b*) *Bar charts.*
- (*i*) Simple bar charts.
- (*ii*) Component bar charts (actuals).
- (*iii*) Percentage component bar charts.
- (*iv*) Multiple bar charts.

(*c*) *Pie charts.*

Each will be considered in turn and illustrative examples will be taken from the data in Table IV.

TABLE IV. DOMICILE OF STUDENTS ATTENDING LITTLE FIELDING RESIDENTIAL COLLEGE, 1972–76

		Domicile in England		
Year	*Total students*	*South*	*Midlands*	*North*
1972	115	81	32	2
1973	135	85	46	4
1974	161	90	62	9
1975	150	77	59	14
1976	215	97	90	28
1976:				
Urban	160	76	62	22
Rural	55	21	28	6

Source: *L.F.R.C. records*

PICTORIAL REPRESENTATION

1. Pictograms. This form of presentation involves the use of pictures to represent data. There are two kinds of pictogram:

(*a*) Those in which the same picture, always the same size, is shown repeatedly—the value of a figure represented being indicated by *the number of pictures shown* (*see* Fig. 8(*a*)).

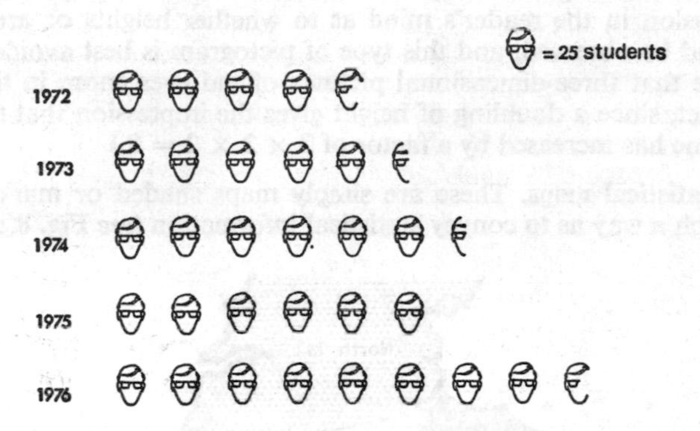

FIG. 8(*a*) *Diagrammatic representation.* Pictogram.

(*b*) Those in which the pictures change in size—the value of a figure represented being indicated by *the size of the picture shown* (*see* Fig. 8(*b*)).

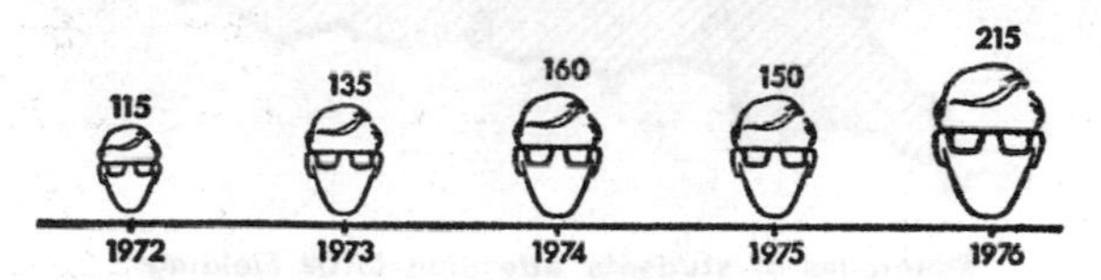

Number of students attending Little Fielding Residential College
1972–76

FIG. 8(*b*) *Diagrammatic representation.* Misleading pictogram.

Type (*b*) is definitely not recommended as it can be very misleading. If the figure being represented doubles, for example, such an increase would probably be shown by doubling the height of the picture. However, if the height is doubled the width must also be doubled to keep the picture correctly proportioned and this results in the area increasing by a factor of 4. To the eye, then, it may well appear that the figure has quadrupled!

Sometimes an attempt is made to overcome the problem by simply doubling the area, but inevitably there is always some confusion in the reader's mind as to whether heights or areas should be observed, and this type of pictogram is best avoided. (Note that three-dimensional pictures offend even more in this respect, since a doubling of height gives the impression that the volume has increased by a factor of $2 \times 2 \times 2 = 8$.)

2. Statistical maps. These are simply maps shaded or marked in such a way as to convey statistical information (*see* Fig. 8(*c*)).

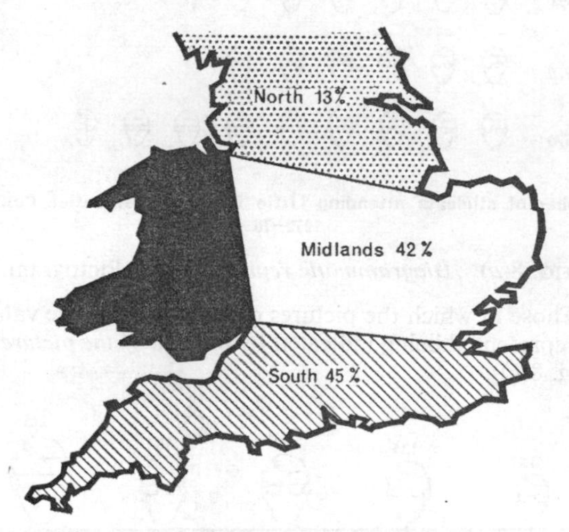

Domiciles of students attending Little Fielding Residential College 1976

FIG. 8(*c*) *Diagrammatic representation.* Statistical map.

3. Uses of pictograms and statistical maps. These two types of diagram are very elementary forms of visual representation, but they can be more informative and more effective than other methods for presenting data to the general public, who, by and large, lack the understanding and interest demanded by the less attractive forms of representation.

Statistical maps are particularly effective in bringing out the geographical pattern that may lie concealed in the data. For instance, it is clear from Fig. 8(*c*) that the further north one goes the fewer the students who attend Little Fielding Residential College.

BAR CHARTS

4. Simple bar charts. In simple bar charts, data are represented by a series of bars: the height (or length) of each bar indicating the size of the figure represented (*see* Fig. 8(*d*)).

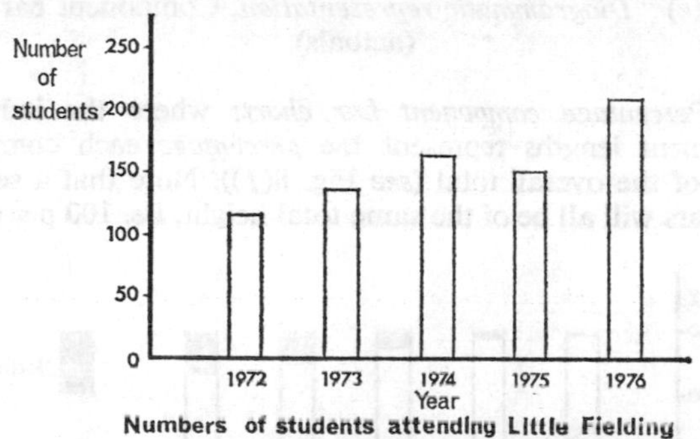

FIG. 8(*d*) *Diagrammatic representation.* Simple bar chart.

Since bar charts are similar to graphs, virtually the same principles of construction apply (*see* V)—though note there should *never* be a "break to zero" in bar charts.

5. Component bar charts. These are like ordinary bar charts except that the bars are subdivided into component parts. This sort of chart is constructed when each total figure is built up from two or more component figures. They can be of two kinds:

(*a*) *Component bar chart* (*actuals*): when the overall height of the bar and the individual component lengths represent *actual* figures (*see* Fig. 8(*e*)).

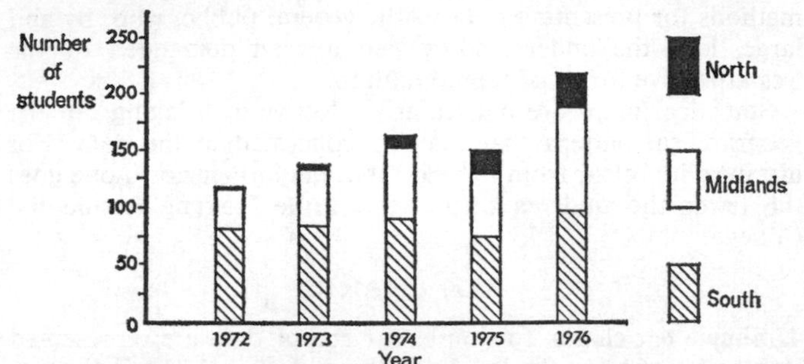

FIG. 8(*e*) *Diagrammatic representation.* Component bar chart (actuals).

(*b*) *Percentage component bar chart:* where the individual component lengths represent the *percentage* each component forms of the overall total (*see* Fig. 8(*f*)). Note that a series of such bars will all be of the same total height, i.e. 100 per cent.

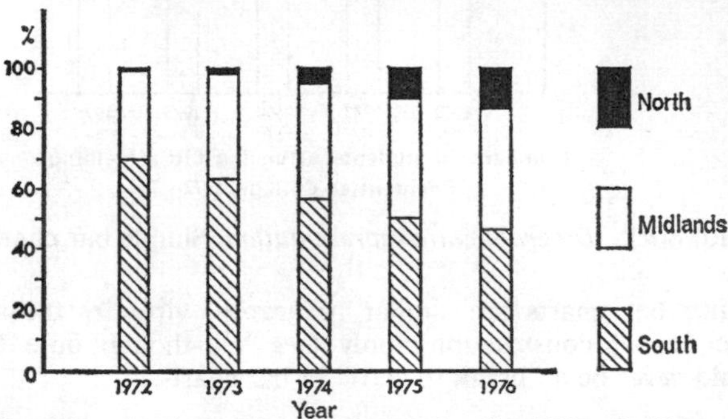

FIG. 8(*f*) *Diagrammatic representation.* Percentage component bar chart.

6. Multiple bar charts. In this type of chart the component figures are shown as *separate bars adjoining each other.* The height of each bar represents the actual value of the component figure (*see* Fig. 8(*g*)).

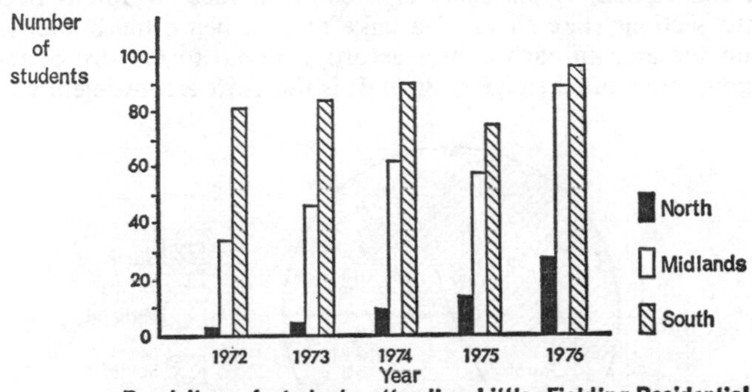

FIG. 8(*g*) *Diagrammatic representation.* Multiple bar chart.

7. Uses of bar charts. Bar charts are very useful for depicting in a simple manner a *series of changes* in major figures. They are usually preferable to pictograms because:

(*a*) they are easier to construct,
(*b*) they can depict data more accurately,
(*c*) they can be used to indicate the sizes of component figures.

8. Which type should be used? The appropriate type of bar chart can be decided upon as follows:

(*a*) *Simple bar charts* where changes in totals only are required.

(*b*) *Component bar charts* (*actuals*) where changes in totals *and* an indication of the size of each component figure are required.

(*c*) *Percentage component bar charts* where changes in the *relative size only* of component figures are required.

(*d*) *Multiple bar charts* where changes in the actual values of the component figures *only* are required, and the overall total is of no importance.

Component and multiple bar charts can only be used, however, when there are not more than three or four components. More components make the charts too complex to enable worthwhile visual impressions to be gained. When a large number of components have to be shown, a pie chart is more suitable.

PIE CHARTS

9. Description. A *pie chart* is a circle divided by radial lines into sections (like slices of a cake or pie; hence the name) so that the area of each section is proportional to the size of the figure represented (*see* Fig. 8(*h*)). It is therefore a convenient way

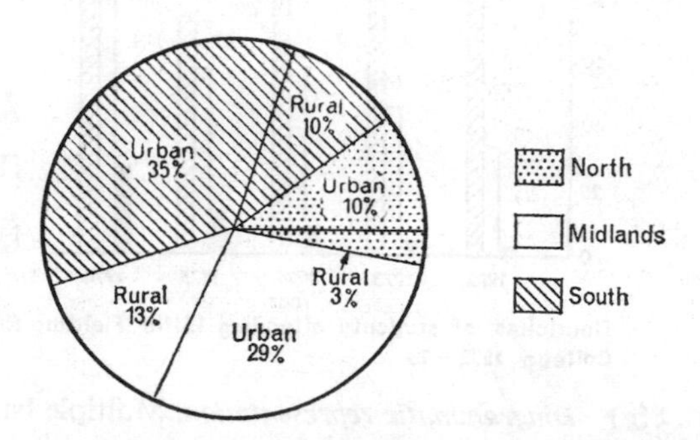

Domiciles of students attending Little Fielding Residential College 1976

FIG. 8(*h*) *Diagrammatic representation.* Pie chart.

of showing the sizes of component figures in proportion to each other and to the overall total.

10. Pie chart construction. Geometrically it can be proved that if the section angles at the centre of the "pie" are in the same proportions as the figures to be illustrated, then the areas of the sections must also be in proportion. To construct a pie chart, then, it is only necessary to construct angles at the centre of the "pie" in proportion to the figures concerned. Thus the 76 southern urban students in Table IV constituted 35 per cent of the total number of students and hence the angle of their section in Fig. 8(*h*) was 35 per cent of 360° = 126° (remember, the total number of degrees at the centre of a circle is 360). Similarly, the 10 per cent northern urban students' section was drawn by constructing an angle of 10 per cent of 360° = 36°.

11. Uses. A pie chart is particularly useful where it is desired to show the relative proportions of the figures that go to make up a single overall total. Unlike bar charts, it is not restricted to three or four component figures only—although its effectiveness tends to diminish above seven or eight.

Pie charts cannot be used effectively where a *series* of figures is involved, as a number of different pie charts are not easy to compare. Again, note that changes in the overall totals should not be shown by changing the size of the "pie," for the same reason as one should not change the size of pictograms.

OTHER DIAGRAMMATIC FORMS

Many other diagrammatic forms can be used to present data visually. One interesting example is a wind rose, an illustration of which is given in Fig. 9. This diagram is drawn on a map and

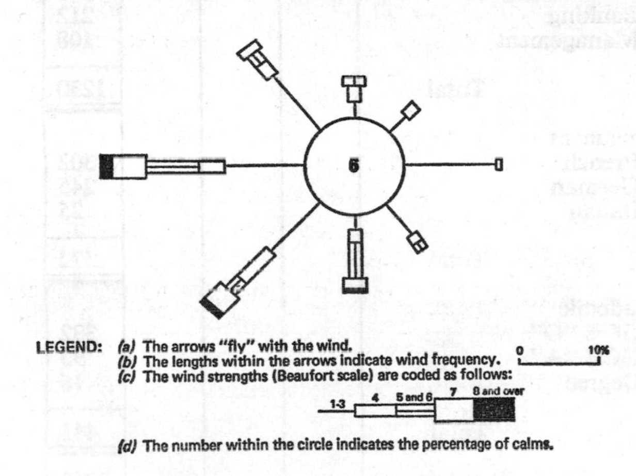

FIG. 9 *Wind rose diagram.*

indicates, by means of carefully constructed "arrows," the pattern of wind directions and speeds at the point where the rose is located.

Indeed, it would appear that the only limitation to the various forms of diagrammatic presentation of data is that set by human ingenuity.

PROGRESS TEST 8

1. Depict the data in the part table below in the form of*:

(*a*) A simple bar chart.
(*b*) A component (actuals) bar chart.
(*c*) A percentage component bar chart.
(*d*) A multiple bar chart.
(*e*) A pie chart.

SUBJECTS STUDIED AT UPPER GUMTRY COMMERCIAL COLLEGE

Subject	*Number of students*				
	1972	1973	1974	1975	1976
Professional					
Accountancy				910	
Banking				212	
Management				108	
Total				1230	
Languages					
French				302	
German				245	
Italian				25	
Total				572	
Academic					
GCE "O"				332	
GCE "A"				93	
Degree				16	
Total				441	
Total all students				2243	

PART THREE

FREQUENCY DISTRIBUTIONS

CHAPTER IX

Frequency Distributions

Look at Table V. What information can be gleaned from this mass of figures?

TABLE V. RAW DATA

Distances (km) recorded by 120 salesmen in the course of one week

482	502	466	408	486	440
470	447	413	451	410	430
469	438	452	459	455	473
423	436	412	403	493	436
471	498	450	421	482	440
442	474	407	448	444	485
505	515	500	462	460	476
472	454	451	438	457	446
453	453	508	475	418	465
450	447	477	436	464	453
415	511	430	457	490	447
433	416	419	460	428	434
420	443	456	432	425	497
459	449	439	509	483	502
424	421	413	441	458	438
444	445	435	468	430	442
455	452	479	481	468	435
462	478	463	498	494	489
495	407	462	432	424	451
426	433	474	431	471	488

Certainly it can be seen that most of the distances are in the 400's, though an occasional figure in the 500's is noticeable. But, once more, the limitations of the human mind make it impossible to detect whether there is any pattern in the figures. Are they spread out evenly, for example, or are there points of concentration? To answer such questions, statistical techniques can be used to process such a mass of figures relating to a single variable so that their significance can be grasped. This part of the book considers the various techniques involved.

ARRAYS AND UNGROUPED FREQUENCY DISTRIBUTIONS

1. Array. The distances in Table V were obtained by simply listing the figures as they were given by the salesmen. Such a collection of figures recorded as they are received is referred to as *raw data.*

The first obvious step to be taken in making the raw data more meaningful is to re-list the figures in order of size, i.e. re-arrange them so that they run from the lowest to the highest. Such a list of figures is called an *array*. An array of the distances in Table V is shown in Table VI.

TABLE VI. ARRAY OF THE RAW DATA IN TABLE V

Distances (km) recorded by 120 salesmen in the course of one week

403	428	440	452	465	483
407	430	441	453	466	485
407	430	442	453	468	486
408	430	442	453	468	488
410	431	443	454	469	489
412	432	444	455	470	490
413	432	444	455	471	493
413	433	445	456	471	494
415	433	446	457	472	495
416	434	447	457	473	497
418	435	447	458	474	498
419	435	447	459	474	498
420	436	448	459	475	500
421	436	449	460	476	502
421	436	450	460	477	502
423	438	450	462	478	505
424	438	451	462	479	508
424	438	451	462	481	509
425	439	451	463	482	511
426	440	452	464	482	515

2. Ungrouped frequency distributions. An examination of the array in Table VI suggests a further simplification. Since some figures repeat (e.g. 407) it would clearly simplify the list if each figure were listed once and the number of times it occurred written alongside, as in Table VII.

TABLE VII. UNGROUPED FREQUENCY DISTRIBUTION CONSTRUCTED FROM THE ARRAY IN TABLE VI

Distances (km) recorded by 120 salesmen in the course of one week

Distance	Frequency	Distance	Frequency	Distance	Frequency	Distance	Frequency
403	1	434	1	456	1	479	1
407	2	435	2	457	2	481	1
408	1	436	3	458	1	482	2
410	1	438	3	459	2	483	1
412	1	439	1	460	2	485	1
413	2	440	2	462	3	486	1
415	1	441	1	463	1	488	1
416	1	442	2	464	1	489	1
418	1	443	1	465	1	490	1
419	1	444	2	466	1	493	1
420	1	445	1	468	2	494	1
421	2	446	1	469	1	495	1
423	1	447	3	470	1	497	1
424	2	448	1	471	2	498	2
425	1	449	1	472	1	500	1
426	1	450	2	473	1	502	2
428	1	451	3	474	2	505	1
430	3	452	2	475	1	508	1
431	1	453	3	476	1	509	1
432	2	454	1	477	1	511	1
433	2	455	2	478	1	515	1

Total frequency (Σf) = 120

In statistics the number of occurrences is called the *frequency* (and symbolised as f), and what we have in Table VII is called an *ungrouped frequency distribution* ("ungrouped" simply distinguishes it from the grouped distribution discussed in 3). An ungrouped frequency distribution, then, is a list of figures occurring in the raw data, together with the frequency of each figure.

Note that the sum of the frequencies (Σf) must equal the total number of items making up the raw data.

GROUPED FREQUENCY DISTRIBUTIONS

3. Description. While Table VII (the ungrouped frequency distribution) is an improvement on the array, there are still too many figures for the mind to be able to grasp the information effectively. Consequently it must be simplified even more. This can be done by *grouping* the figures. For example, if all the distances between 400 and under 420 are grouped together there are 12 occurrences in the group, i.e. $f = 12$. These groups are called *classes*. A list of such classes together with their frequencies is called a *grouped frequency distribution.* Thus the information in Table VII can be converted into the grouped frequency distribution shown in Table VIII.

TABLE VIII. GROUPED FREQUENCY DISTRIBUTION

Distances (km) recorded by 120 salesmen in the course of one week

Distance (km)	Frequency (f)
400–under 420	12
420–under 440	27
440–under 460	34
460–under 480	24
480–under 500	15
500–under 520	8
	120

4. Effect of grouping. As a result of grouping, it is possible to detect a pattern in the figures. For instance, the distances in Table VIII cluster around the "440–under 460" class. However, it is important to realise that, although it brings out the pattern, *such grouping does result in loss of information.* For example, the total frequency in the "400–under 420" class is known to be 12, but there is no longer any information as to *where in the class* these 12 occurrences lie. Increased significance has been bought at the cost of loss of information. The exchange is well worth while, but it means that calculations made from a grouped frequency distribution cannot be exact, and consequently excessive accuracy can only result in spurious accuracy.

5. Class limits. *Class limits* are the extreme boundaries of a class. Care has to be taken in defining the class limits, otherwise there may be overlapping of classes or gaps between classes. Imprecision here is a common fault. Given, say, classes of "400–420" and "420–440" kilometres, in which of the two would a distance of 420 be recorded? Obviously, it could be either. Conversely, if we were told that the higher class was "421–440," a distance of 420½ would appear to fit into neither of them. For this reason the classes in Table VIII are stated as "400–under 420," etc. The class limits of the first class are exactly 400 kilometres at the lower end, and *right up to, but not including* 420 kilometres at the upper end (i.e. 419.9999 . . .). A well-designed frequency distribution will ensure that there is neither overlapping of classes nor gaps between them.

6. Stated and true class limits. The class limits actually given in a grouped frequency distribution are called the *stated limits.* Now the stated limits are not necessarily the *true limits.* For example, if a distribution was headed "Age last birthday" and the first class was "10–19 years," i.e. stated limits of "10 years" and "19 years," we would know from common sense that this meant the *true* limits were "10 years exactly" and "19 years, 364 days." (This must be since anyone who had passed his 19th birthday but who had not reached his 20th birthday was "19 last birthday" and so would be recorded in the "10–19 years" class.) In this sort of situation it is very important for students to remember that one should *always use the true limits and never the stated ones*—and that the true limits are found by using common sense. There will, of course, be many occasions when the stated limits are also the true limits but one should always check first before using the stated limits in any statistical work.

7. Discrete and continuous data. At this juncture students should be warned that data exist in one of two forms, discrete or contінous.

Discrete data are data that increase *in jumps.* For instance, if the data relate to the numbers of children in families then the figures recorded will be 0, 1, 2, 3, or 4, etc. 1½ or 2¼ children are impossible figures. In other words, fractions of a unit are impossible and the data increase in jumps—from 0 to 1 from 1 to 2, etc. (Note, however, the units themselves can be fractions —e.g. ½p.)

Continuous data on the other hand are data that can increase *continuously.* If, say, kilometres travelled are being investigated the figures recorded could end in any fraction imaginable, e.g. 425.001, 425.634, 425.999, etc.

Now it is very important to note that whether data is discrete or continuous *depends solely upon the real nature of the data and not upon how it is collected.* Thus the salesman's mileages discussed so far are continuous data since any fraction of a mile can occur in reality although a glance at the raw data (Table V) reveals no fractions at all, i.e. they were recorded as if they were discrete data. To repeat, it is not how the data are recorded that matters but only how it occurs in reality, and on this basis alone one decides if data are continuous or discrete.

8. Mathematical limits. Let us now consider the true class limits of a distribution of discrete data. Say, for example, the *stated* limits of a class are "5–under 10." Now if the data is discrete obviously there can be no occurrences above 9 and below 10, and so using the *true* limits the class is "5–9." It also follows that the lower limit of the next class is 10.

Now according to the rule given in **6** we should use these true limits in our statistical calculations, but for purely mathematical reasons it is very awkward working with a gap between classes—albeit in nature there really is a gap. Consequently we close the gap by extending the true limits of each class by ½ a unit—i.e. our "5–9" class now becomes "4½–9½" and our next class starts at 9½. These extended limits are called *mathematical limits*, and *if our data is discrete we must always work with the mathematical limits.* (Note, incidentally, that a class with true limits "0–4" will have mathematical limits "−½–4½." Do not let this −½ bother you. It is purely a mathematical limit and you can be assured that all comes out well with figures computed using this negative limit.)

Note that all the rules involving class limits are summarised diagrammatically in Fig. 10.

9. Class limits and the form of raw data. In determining the true class limits attention should be paid to the form of the raw data. For instance, in the raw data of salesmen's distances shown in Table V there are no fractions of a kilometre. Yet it is inconceivable that the distances travelled by so many salesmen could all be exact kilometres. There must have been some rounding.

Now, if the figures were rounded to the nearest kilometre,

Example—see **8.**

Stated class limits 5–under 10
(Limits stated in the distribution)
Never work with these as such
Use them only to obtain ——→ *True class limits* 5–9

If data are *continuous* work with these limits
If data are *discrete*— use to obtain ——→ *Mathematical class limits* . . . . $4\frac{1}{2}$–$9\frac{1}{2}$

These are found by extending the true limits outwards by $\frac{1}{2}$ unit.
Always work with these limits when data is discrete.

FIG. 10. *Class limit rules.*

It is important when working with frequency distributions to use the correct class limits. This figure summarises the rules given in **5** to **8**.

419.75 kilometres, for example, would be recorded as 420 kilometres. That means it would be grouped in the "420–under 440" class in Table VIII. *But clearly such a distance should be in the "400–under 420" class.* Therefore Table VIII would be incorrectly constructed. Its construction can only be correct if the raw data were recorded on the basis of the number of *completed* kilometres.

In practice such fineness would usually be unnecessary, but this example demonstrates that some care is needed in determining class limits if grouped frequency distributions are to be correctly constructed.

10. Class interval. This is the width of the class—in other words, the difference between the class limits (the true or mathematical class limits, of course, depending upon whether the data is discrete or continuous). If the class intervals of all the classes are equal, the distribution is said to be an *equal class interval distribution.* In Table VIII the class interval is 20 kilometres for all classes and the distribution is therefore an equal class interval distribution.

11. Unequal class intervals. Some sets of figures are such that, if equal class intervals were taken, very few classes would contain nearly all the occurrences whilst the majority would be virtually empty, e.g. a distribution of annual salaries with class intervals of £1000. In cases like this, it is better to use unequal class intervals. They should be chosen so that the over-full classes are subdivided and the near-empty ones grouped together, e.g. 0–under £1500, £1500–under £2500, £2500–under £3000, £3000–under £4000, £4000–under £5000, £5000–under £7500, £7500–under £10 000, etc.

12. Open-ended classes. If the first class in a distribution is stated simply as "Under . . ." (e.g. "Under 400 kilometres") or the last is stated as "Over . . ." (e.g. "Over 500 kilometres"), such classes are termed *open-ended,* i.e. one end is open and goes on indefinitely. They are used to collect together the few extreme items whose values extend way beyond the main body of the distribution.

The class interval of an open-ended class is by convention deemed to be the same as that of the class immediately adjoining it. In well-designed distributions, open-ended classes have very low frequencies and so the error that may arise from using the convention is not important.

13. Choice of classes. The construction of a grouped frequency distribution always involves making a decision as to what classes shall be used. The choice will depend on individual circumstances, but the following suggestions should be borne in mind:

(*a*) Classes should be between ten and twenty in number.

NOTE: For the sake of simplicity the number of classes in examples in this book will be kept very small.

(*b*) Class intervals should be equal wherever practicable.

(*c*) Class intervals of 5, 10 or multiples of 10 are more convenient than other intervals such as 7 or 11.

(*d*) Classes should be chosen so that occurrences within the classes tend to balance around the mid-points of the classes. It would be unwise to have a class of (say) "£2000–under £2100" in a salary distribution, since salaries at this level are often in round £100's, so most of the occurrences would be concentrated at £2000, i.e. the extreme end of the class. This is unsatisfactory, since later theory makes the assumption that the average of the occurrences in a class lies at the mid-point of the class.

14. Direct construction of a grouped frequency distribution. Once the raw data have been recorded (as in Table V) we may wish to construct a grouped frequency distribution directly, without going through the intermediate steps of an array and an ungrouped frequency distribution. The steps are as follows:

(*a*) Pick out the highest and lowest figures (Table V: 403 and 515) and on the basis of these decide upon and list the classes.

(*b*) Take each figure in the raw data and insert a tally mark (|) against the class into which it falls (*see* Table IX). Note that every fifth tally mark is scored diagonally across the previous four. This simplifies the totalling at the end.

(*c*) Total the tally marks to find the frequency of each class.

The distribution is now complete, though the class frequencies should be added up and the total checked to see that it corresponds with the total number of occurrences in the raw data.

TABLE IX. DIRECT CONSTRUCTION OF TABLE VIII FROM RAW DATA
(*see* Table V)

Class	*Check marks*	*Frequencies*
400–under 420	𝍸 𝍸 ||	12
420– „ 440	𝍸 𝍸 𝍸 𝍸 𝍸 ||	27
440– „ 460	𝍸 𝍸 𝍸 𝍸 𝍸 𝍸 ||||	34
460– „ 480	𝍸 𝍸 𝍸 𝍸 ||||	24
480– „ 500	𝍸 𝍸 𝍸	15
500– „ 520	𝍸 |||	8
		120

PROGRESS TEST 9

(*Answers in Appendix VI*)

1. Suggest classes for insertion into the "Classes" column of grouped frequency distributions compiled from raw data relating to:

(*a*) A survey of the ages of adults in a city. Questionnaires were sent to all people of 20 years and over, asking them to state their present age in years.

(*b*) A survey of the number of extractions made on a specific day by a group of dentists, the numbers ranging between 4 and 32.

(*c*) Incomes per annum of all full-time employed adults in a town (recorded to the nearest £).

State the exact class limits of the second class chosen by you for each of the distributions.

2. Reconstruct the grouped frequency distribution for Table V, using the same class interval but starting with the first class at 390 kilometres.

CHAPTER X

Graphing Frequency Distributions

HISTOGRAMS

1. Description. A *histogram* is the graph of a frequency distribution. It is constructed on the basis of the following principles:

(*a*) The horizontal axis is a continuous scale running from one extreme end of the distribution to the other. This means that this axis is *exactly the same as any ordinary axis on a graph.* It should be labelled with the name of the variable and the units of measurement.

(*b*) For each class in the distribution a vertical rectangle is drawn with:

(*i*) its base on the horizontal axis extending from one class limit of the class to the other class limit;

(*ii*) its *area proportional to the frequency in the class* (i.e. if one class has a frequency twice that of another, then its rectangle will be twice the area of the other).

NOTE

(*i*) The class limits at the ends of the rectangle base are the true limits in the case of continuous data and the mathematical limits in the case of discrete data. (This means, incidentally, that there will *never be any gaps* between the histogram rectangles.)

(*ii*) If the distribution is an *equal class interval distribution* (the more usual case) then the bases of all the rectangles will be the same lengths. This means that to obtain areas proportional to frequencies the heights must be drawn proportional to the frequencies (i.e. a class having twice the frequency of another will have a rectangle twice the height), and in this particular case the vertical axis of the graph can be labelled "Frequency."

The histogram for Table VIII is shown in Fig. 11. As this is an equal class interval distribution the heights of the rectangles are in this case drawn proportional to the frequencies of the classes.

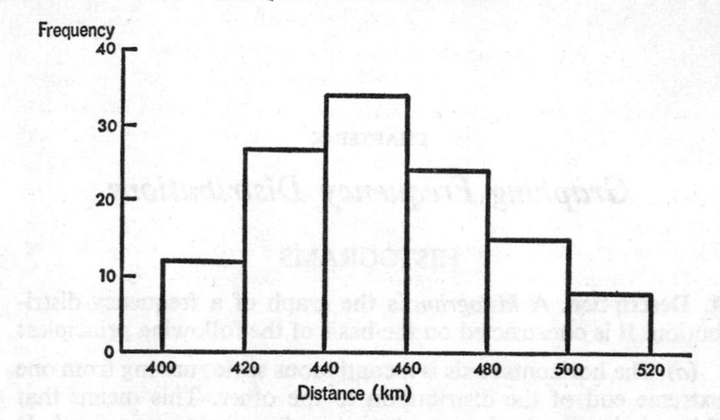

FIG. 11 *Histogram of data from Table VIII.*

2. Distributions with unequal class intervals. If, in Table VIII, classes "400–under 420" and "420–under 440" were merged to give a single class, "400–under 440," the combined frequency

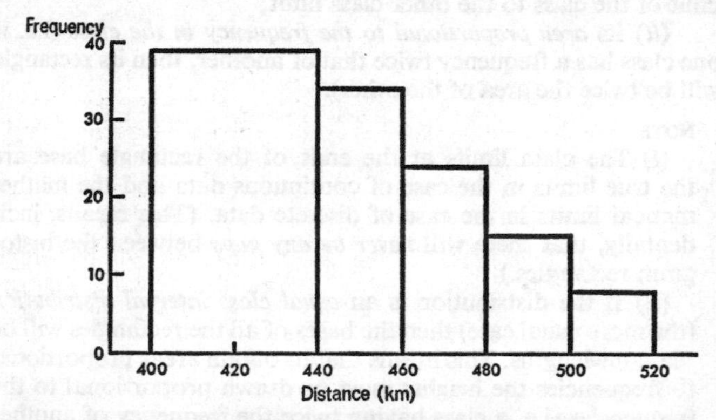

FIG. 12 *Histogram with unequal class intervals.* Incorrect construction.

would be 39. Now suppose we construct a histogram and that the rectangle for this group is drawn two class intervals wide, with a frequency of 39. The resulting graph (Fig. 12) is seriously in error. Compared with the original histogram (Fig. 11), it is obvious that the area of the combined classes is much greater than the combined areas of the two separate classes.

As it is essential that the total areas should be the same, an adjustment must be made. This adjustment is quite simple, for since the "400–under 440" class has a class interval double that of the other classes, its histogram rectangle has a width double that of the other rectangles. Now, Area = Width × Height, so a doubling of the width can be adjusted for by halving the height. Thus the correct histogram is drawn with a frequency of $19\frac{1}{2}$ (i.e. half of 39) for the "400–under 440" class (Fig. 13).

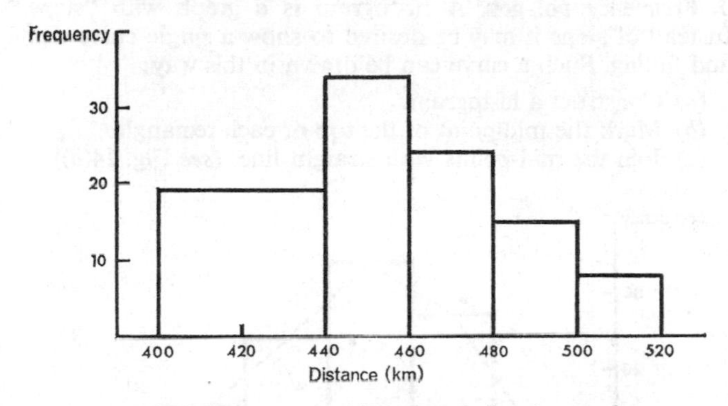

FIG. 13 *Histogram with unequal class intervals.* Correct construction.

NOTE

(*i*) Compared with the original histogram, this average height still does not look quite correct. This distortion arises through the loss of information as a result of further grouping.

(*ii*) Strictly speaking, the vertical axis is no longer "Frequency" but "Frequency density"—i.e. it measures how densely packed the occurrences are within each class (e.g. although class 400–under 440 has 39 occurrences these are spread over a 40-kilometre class and hence are "packed" together at an average density of less than one per kilometre, while the class 440–under 460 has 34 occurrences spread over 20 kilometres—i.e. on average 1.7 occurrences per kilometre).

Similarly, if a frequency distribution has a class with an

interval three times the normal, the frequency of that class must be divided by three.

NOTE: Occasionally, distributions arise with no normal class interval. In such cases the smallest class interval should be assumed to be "normal."

FREQUENCY POLYGONS AND FREQUENCY CURVES

3. Frequency polygon. A histogram is a graph with "steps." Instead of steps it may be desired to show a single curve rising and falling. Such a curve can be drawn in this way:

(*a*) Construct a histogram.
(*b*) Mark the mid-point of the top of each rectangle.
(*c*) Join the mid-points with straight lines (*see* Fig. 14(*a*)).

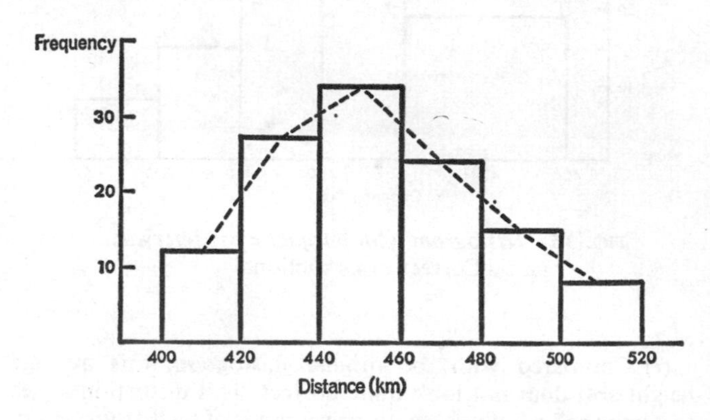

FIG. 14(*a*) *Frequency polygon of data from Table VIII.* Step (*c*).

This results in a single curve, but it does not reach the horizontal axis of the graph; it stops at the mid-points of the outermost steps. It is desirable that the curve should be "anchored" to the horizontal axis so that it encloses an area equal to that of the histogram. The final step, then, is:

(*d*) Extend the curve downwards at both ends so that it cuts the axis at points which are *half a class interval beyond the outside limits of the end classes* (*see* Fig. 14(*b*)). This should be done even if it means that the curve crosses the vertical axis and ends in the minus part of the graph.

The resultant curve gives us a figure known as a *frequency polygon.* Figure 14(*b*) is in fact a frequency polygon of Table VIII superimposed on the histogram.

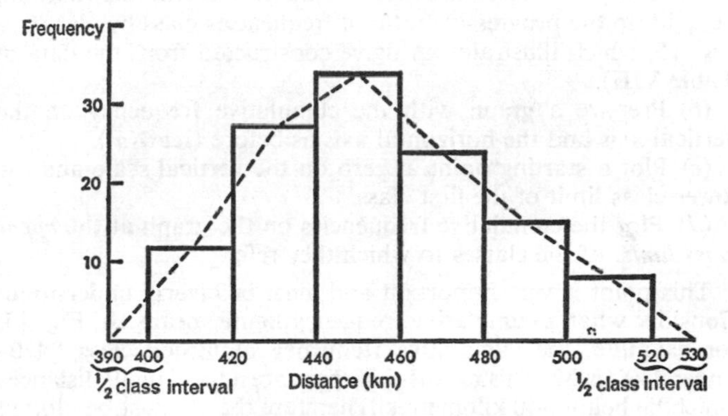

FIG. 14(*b*) *Frequency polygon of data from Table VIII.* Step (*d*).

4. Area of a frequency polygon. It will be appreciated that in drawing a frequency polygon the corner of each rectangle cut off by the polygon is equal in area to the triangle added between the point where the polygon line emerges from one rectangle and the mid-point of the next. Thus the area of the frequency polygon is exactly equal to the area of the histogram. This of course also explains why the polygon is extended *half* a class outwards when it is brought down to meet the horizontal axis in step (*d*) above.

5. Frequency curve. If the frequency polygon is smoothed so that there are no sharp points, it is known as a *frequency curve.* Normally, frequency curves should only be constructed when there are a large number of classes and very small intervals—when, in fact, the frequency polygon is virtually a smooth curve anyway. A frequency curve constructed under other circumstances tends to be inaccurate.

Again, the area under a frequency curve must be equal to the area of the histogram on which it is based.

OGIVES

6. Definition. *Ogive* is the name given to the curve obtained when the *cumulative* frequencies of a distribution are graphed. It is also called a *cumulative frequency curve.*

7. Ogive construction. To construct an ogive:

(*a*) Compute the cumulative frequencies of the distribution, i.e. add up the progressive total of frequencies class by class (*see* Fig. 15, which illustrates an ogive constructed from the data in Table VIII).

(*b*) Prepare a graph with the cumulative frequency on the vertical axis and the horizontal axis as before (*see* **1**(*a*)).

(*c*) Plot a starting point at zero on the vertical scale and the lower class limit of the first class.

(*d*) Plot the cumulative frequencies on the graph at the *upper class limits* of the classes to which they refer.

This point is very important and must be clearly understood. Consider what a cumulative frequency figure means. In Fig. 15, for example, the cumulative frequency alongside class "420–under 440" is 39. This means that there are a total of 39 distances which lie below 440 kilometres. Therefore the 39 must be plotted just below the 440 kilometre point (for practical graphing purposes, at 440). Similarly, the 73 relating to the "440–under 460" class means a total of 73 distances below 460 kilometres and the 73 must therefore be plotted just below the 460 kilometre point. In other words—to repeat—when constructing an ogive, each cumulative frequency must be plotted at the upper class limit of each class.

(*e*) Join all the points.

8. "Less than" curves. The curve we obtain in this manner is called a "less than" curve because if one takes any point along it, the reading of that point on the cumulative frequency axis gives the number of frequencies that have a value "less than" the reading of that point on the horizontal axis. For instance, in Fig. 15 the curve passes through the point representing a "cumulative frequency of 100" and "484 kilometres." This is read as "100 of the distances are *less than* 484 kilometres." Similarly, at a lower point it can be seen a cumulative frequency of 60 is related to 452 kilometres. This means that 60 distances are of *less than* 452 kilometres.

Class	Frequency	Cumulative frequency	Cumulative percentage (see X, 10)
400–under 420	12	12	10
420– ,, 440	27	39	32.5
440– ,, 460	34	73	61
460– ,, 480	24	97	81
480– ,, 500	15	112	93
500– ,, 520	8	120	100
	120		

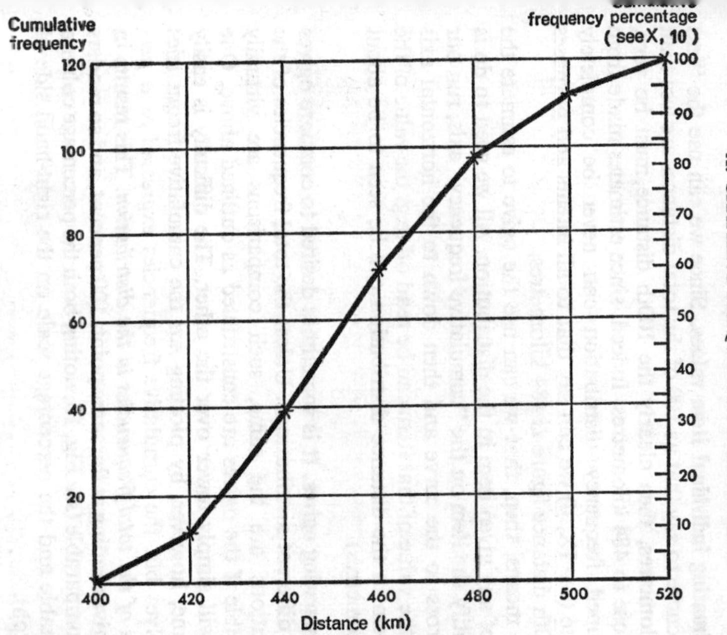

FIG. 15 *Ogive of data in Table VIII ("less than" curve).*
The table alongside the graph shows the cumulative frequency distribution of the data in Table VIII.

9. Estimating individual item values. Since we can use the "less than" curve to say that as 100 of the total distances are less than 484 kilometres, then clearly the 100th distance must be very, very close to 484 kilometres. Indeed, since estimates made from a grouped frequency distribution can never be completely accurate (*see* IX, 4) we can say that to all intents and purposes the 100th distance figure *is* 484 kilometres.

This means, then, that we can use the ogive to estimate the value of any given item in the distribution. All we need to do is to identify the item on the "cumulative frequency" axis, run our eye across to the curve and then down to the horizontal axis where the value of that item can be read off (e.g. the value of the 20th item in the distance distribution can be seen to be about 425 kilometres).

10. Comparing ogives. It is sometimes desired to compare ogives of two different distributions. Unless the total frequencies of the distributions are the same, such comparisons are virtually impossible if the ogives are constructed as outlined above. One ogive will simply tower over the other. The difficulty is easily overcome, however, by plotting *not* the cumulative frequencies themselves but the cumulative frequencies *expressed as a percentage of the total frequencies in the distribution.* This results in both ogives ending at the same point, 100 per cent, and so renders them comparable (*see* Fig. 15, noting both the percentage column in the table and the percentage scale on the right-hand side of the graph).

11. Smoothing ogives. The ogives discussed so far have consisted of a series of straight lines. Normally it is considered dangerous to smooth curves relating to frequency distributions, but in the case of ogives, providing the smoothed curve passes through all the plotted points, such smoothing almost always improves the graph. The reason is that occurrences are not usually spread equally throughout a class but tend to cluster in one or the other half of the class. Smoothing automatically makes some allowance for this tendency in the case of ogives and therefore improves the accuracy of the curve. Hence ogives are often smoothed.

PROGRESS TEST 10

(*Answers in Appendix VI*)

1. Construct histograms for the following data and superimpose frequency polygons on them*.

MARKS SCORED IN I.Q. TESTS BY PUPILS AT TWO DIFFERENT SCHOOLS

	Number of pupils	
I.Q. marks	*School A*	*School B*
75–under 85	15	43
85– " 95	25	99
95– " 105	40	54
105– " 115	108	40
115– " 125	92	14
125 and over	20	0
	300	250

2. Construct (*a*) a histogram, (*b*) a smooth ogive ("less than" curve), for the following data:

In a certain examination, 12 candidates obtained fewer than 10 marks; 25 obtained 10 to under 25 marks; 51 obtained 25 to under 40 marks; 48 obtained 40 to under 50 marks; 46 obtained 50 to under 60 marks; 54 obtained 60 to under 80 marks, and only 8 obtained 80 marks or more. Marks were out of 100.

3. From Fig. 15 determine:

(*a*) the number of distances recorded below 470 kilometres;

(*b*) the range of distances that were recorded by the lowest 25 per cent of the salesmen.

CHAPTER XI

Fractiles

FRACTILES AND MEDIANS

1. Definition. A *fractile* is the value of the item which is a given fraction of the way through a distribution.

Assume we had a distribution in the form of a simple array of data and we were asked to give the value of the item one-third of the way through this distribution. The answer could easily be given by starting at the beginning of the distribution, counting until one came to the item one-third the way through the distribution, and then reading off the value of that item. This value is the required fractile.

One point needs great emphasis—that is, that a fractile is the *value* of the fractile item. *It is not the fractile item.* Thus, if there were 900 items in the distribution above, the required fractile would be the *value* of the 300th item—it would not be the 300th item itself. Beware, for students frequently make this elementary error.

2. Fractiles and ogives. When we examined ogives we found they could be used for estimating the value of any given item (*see* X, 9). Clearly, then, ogives can be very useful for estimating fractiles—all we need to do is to find which item is the fractile item and use the ogive to read off its value immediately.

Although using an ogive in this way is an excellent method of estimating fractiles, an alternative method is by calculation. Much of the rest of this chapter discusses this alternative method so as to provide the student with two different ways of finding fractiles.

Both, incidentally, give equally correct answers—unless the examination question calls for one or the other the student may select either. In practice, calculation generally proves quicker if finding only one or two fractiles, while the ogive is quicker if more fractiles are required.

3. Median. Obviously, one can take any fraction of a distribution one wants—third, quarter, three-tenths, etc.—and find the corresponding fractile, but the most common fraction is a half. The fractile obtained using half as the required fraction is called the *median*, which can therefore be defined as the value of the middle item. Thus in a distribution of 900 items the median is the *value* of the 450th item.

To find the median then:

(*a*) Arrange the data in an array.
(*b*) Locate the median (middle) item.
(*c*) Read off the value of the median item.

4. Median item in an even-numbered distribution. Strictly speaking, the 450th item referred to above is not the middle item in a distribution of 900 items. There is, in fact, no middle item, the mid-point of the distribution lying between the 450th and the 451st items. (If this is not quite clear, imagine a distribution with only four items. Which is the middle item? Clearly, there is not one, the mid-point of the distribution lying between items 2 and 3.)

The fact that an even-numbered distribution has no single mid-point item often bothers students but in practice there is no difficulty, since in any realistic distribution the values of the middle two items will be virtually the same and either will suffice (if they are not then it would be unwise to use the median at all). Alternatively, the values of the middle two items may be averaged.

5. Median of a grouped frequency distribution. Finding the median of a simple array is easy enough but finding the median of a grouped distribution is rather more difficult. One cannot now read off the value of the median item since in grouping the data the values of individual items have been lost. Nevertheless it is possible to *estimate* the median. The necessary procedure to do this is given below using the distribution in Table VIII for illustration.

First we find the median item. Since there are 120 items in the Table VIII distribution the median item is the 60th.

Now from a cumulative frequency distribution it is possible to determine within which class the median item lies. This class is called the *median class* (note that a grouped frequency distribution is laid out as an *array of classes*). In the case of Table VIII the cumulative frequency distribution is as follows:

Kilometres	*Frequency*	*Cumulative frequency*
400–under 420	12	12
420– „ 440	27	39
440– „ 460	34	73 (median class)
460– „ 480	24	97
480– „ 500	15	112
500– „ 520	8	120

The 60th item lies, therefore, in the "440–under 460" class, and so that is the median class.

The median itself is now estimated by computing the probable location of the median item within the median class. In our example, the 60th item is the 21st item from the bottom of the median class (the cumulative frequency distribution shows that the 39th item just falls in the class preceding the median class, and the 60th item is just 21 more items further on, i.e. 21st from the bottom of the class). Now there are 34 items in the median class and if it is assumed these items are equally spread throughout the class the median item must be 21/34ths of the way through the class. Since the class interval is 20 kilometres, then 21/34ths of this interval is $21/34 \times 20 \simeq 12.4$ kilometres. So the median item lies 12.4 kilometres into the median class and, since this class starts at 440 kilometres, the value of the median item (i.e. the "median") must be approximately $440 + 12.4 = 452.4$ kilometres.

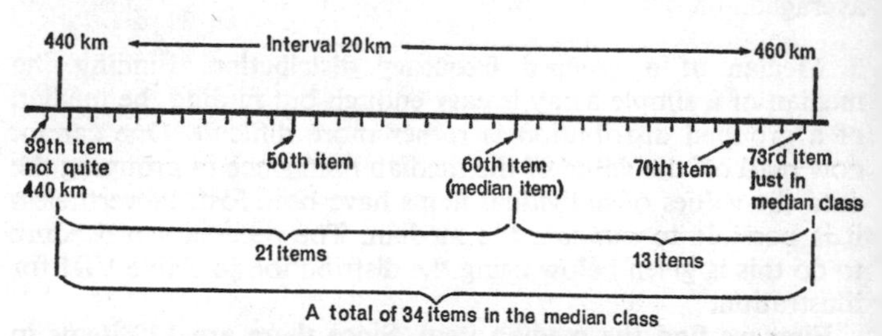

FIG. 16 *Locating the median item within the median class.*

This may all seem rather complicated, but all we are doing is to determine how far the median item lies along the median class interval by means of proportions. Figure 16, showing the position diagrammatically, may help to make things clearer. It can be seen that the median item is 21/34ths of the interval of 20 kilometres, i.e. 12.4 kilometres. Therefore the value of the median item (i.e. the median) is $440 + 12.4 = 452.4$ kilometres.

6. Use of median. The median is a useful statistical measure for as it is the value of the item in the middle of a distribution it means that generally speaking half of all the items have values above the median and half have values below the median. Since a few items may have the same value as the median (particularly if the data is discrete) it is more accurate to say that half of the items have a value *equal to or greater than* the median and half a value *equal to or less than* the median.

7. Estimating fractiles from a grouped frequency distribution. The median is, of course, a fractile and everything that has been said about the procedure for estimating a median from a grouped distribution can be applied to the estimation of any other fractile —i.e. one simply locates the fractile class, finds by proportion how far the fractile item lies in the fractile class, and adds this value to the lower class limit of the fractile class (*see* **13** for summary of procedure). Two points should be noted in estimating fractiles in this manner:

(*a*) Clearly a correct estimate depends upon correctly determining the lower class limit of the fractile class. As usual, the true limits must be used for continuous data and the mathematical limits for discrete data.

(*b*) When a fractile is estimated from a grouped frequency distribution of discrete data the answer frequently contains a fraction of a unit. Since the object of estimating a fractile is to give an actual single value—that of the fractile item—and since such a fraction is impossible in the case of discrete data, it is necessary to round off the answer to the nearest unit. If, for example the data in Table VIII had related to numbers of cars, the median value of 452.4 would have been rounded to 452 cars.

QUARTILES, DECILES AND PERCENTILES

8. Quartiles. Quartiles are the fractiles relating to the one-quarter and three-quarters fractions. The *quartiles*, then, can be defined as the values of the items one-quarter and three-quarters of the way through a distribution. The value of the item one-quarter of the way through the distribution is called the *lower* or *first quartile* (symbolised as Q_1) and the item three-quarters of the way the *upper* or *third quartile* (symbolised as Q_3).

In the case of a simple array finding the quartiles is just a matter of reading off the values of the quartile items, but where a grouped frequency distribution is involved the quartiles must be estimated in the same way as the median.

EXAMPLE (using Table VIII).

In the distribution there are 120 items. The quartile items are, therefore, the 30th and 90th items. By constructing a cumulative frequency distribution, the classes containing the quartiles can be found:

Kilometres	*Frequency*	*Cumulative frequency*
400–under 420	12	12
420– „ 440	27	39 (Q_1 in this class)
440– „ 460	34	73
460– „ 480	24	97 (Q_3 in this class)
480– „ 500	15	112
500– „ 520	8	120

This distribution reveals the classes within which the quartile items lie. The *values* of these items are found in exactly the same way as the value of the median was found, i.e. by assuming the items to be equally spread throughout each class and then computing the value of the required item by means of proportions. Thus, since the first quartile item is the 30th, and since the first 12 items lie in an earlier class, it is the $30 - 12 = 18$th item in its class of 27 items, i.e. it lies 18/27ths of the way into the class. As this class has an interval of 20 kilometres this is equivalent to $18/27 \times 20 = 13.33$ kilometres. Now the quartile class starts at 420 kilometres, so the first quartile $= 420 + 13.33 = 433.33$ kilometres.

Similarly $Q_3 = (90 - 73)/24 \times 20 + 460 = 474.1$ kilometres.

9. Use of quartiles. The median is a valuable statistical measure but, because it is the value of only one item out of a distribution of possibly hundreds, it often needs supplementing by other measures. One of its shortcomings, for instance, is that it gives no indication as to how far on either side of it the other values extend.

To know that the median wage in a department is £76 per week gives us no idea of the actual level of pay received by most of the lower-paid employees, except that it is not above £76. If, however, it is also known that the first quartile is £70 and the third £98, then it is clear that many employees earn a wage close to the median, while some earn much more (a quarter of the employees must earn £98 or more by the very definition of a quartile). Indeed, it is possible to go further. Since a quarter of the distribution lies below the first quartile and a quarter lies above the third, then *half the distribution must lie between the two quartiles.* It can be concluded, then, that half the employees earn between £70 and £98 per week, whilst a quarter earn £70 or less and a quarter earn £98 or more.

From this it can be seen that a considerable amount of information will be contained in a statement that gives only three figures, if they are the median and the two quartiles. For this reason, if it is wished to summarise a large set of figures very briefly, the median and the two quartiles may be chosen as the representative figures that carry the most information.

10. Deciles. *Deciles* are fractiles relating to *tenths* of the way through a distribution. Thus the first decile is the value of the item 1/10 of the way through a distribution (in the case of Table VIII, the 120/10 = 12th item). The second decile is the value of the item 2/10 of the way (in Table VIII the 120 × 2/10 = 24th item) and the, say, ninth decile the value of the item 9/10 of the way (in Table VIII the 108th item). The values of these items are found in exactly the same way as the values of medians and quartiles are found.

EXAMPLE

Find the third decile of the data in Table VIII.

The third decile in the item is 3/10 of the way through the distribution,

i.e. $\frac{3}{10} \times 120 =$ the 36th item.

From the cumulative frequency distribution shown in XI, **8** it can be seen that this item is in the "420–under 440" class and is the 36 − 12 = 24th item in a class of 27 items.

$\therefore$ Value of this item is $\left(\frac{24}{27} \times 20\right) + 420 = 437.8$ kilometres.

$\therefore$ The 3rd decile is 437.8 kilometres.

11. Percentiles. *Percentiles* are fractiles relating to *hundredths* of the way through a distribution. Thus the 85th percentile is the value of the item 85/100ths of the way through the distribution.

EXAMPLE

Find the 85*th percentile of the data in Table VIII.*

The 85th percentile is the $\frac{85}{100} \times 120 = 102$nd item in the array.

This item lies in the "480–under 500" class and its value is $\left(\frac{102 - 97}{15} \times 20\right) + 480 = 486\frac{2}{3}$ kilometres.

12. Use of percentiles. Percentiles are particularly useful as "cut-off" values.

For example, assume that only 22 per cent of boys applying for places in a public school can be accepted and that acceptance is based on I.Q. If the boys' I.Q.s are tested and set out in an array, the 78th (100–22) percentile will indicate the I.Q. level that will separate the successful applicants from the others.

13. Summary—estimating fractiles from a grouped frequency distribution. For convenience we summarise here the procedure and formula for estimating any desired fractile from a grouped frequency distribution.

(*a*) Find the *fractile item* by multiplying the desired fraction by the total number of items in the distribution and then rounding up (e.g. the third quartile item in a distribution of 110 items $= \frac{3}{4} \times 110 = 82\frac{1}{2}$, which is rounded up to give the *83rd item*).

(*b*) Prepare a cumulative frequency distribution from the grouped frequency distribution and locate the *fractile class* (i.e. the class containing the required fractile).

(*c*) Apply the following formula:

$$\text{Required fractile} = \text{Lower class limit of fractile class} + \left(\frac{\text{Fractile item} - \text{cumulative frequency up to lower class limit of fractile class}}{\text{Fractile class frequency}} \times \text{Fractile class interval}\right)$$

PROGRESS TEST 11

(*Answers in Appendix VI*)

1. The following marks were obtained by candidates in an examination (no fractions of a mark were awarded):

Marks	*No. of Candi-dates*	*Marks*	*No. of Candi-dates*	*Marks*	*No. of Candi-dates*
0–5	2	36–40	150	71–75	120
6–10	8	41–45	200	76–80	100
11–15	20	46–50	220	81–85	60
16–20	30	51–55	280	86–90	40
21–25	50	56–60	320	91–95	17
26–30	80	61–65	260	96–100	3
31–35	120	66–70	160		
					2240

(*a*) Find: (*i*) the median, (*ii*) the first quartile, (*iii*) the third quartile, (*iv*) the sixth decile, (*v*) the 42nd percentile.
(*Warning:* Think carefully about class limits.)

(*b*) If the examining body wished to pass only one-third of the candidates, what should the pass mark be?

2. From the ogive in Fig. 15, find the 40th percentile. At what percentiles do (*a*) 500 kilometres; and (*b*) 450 kilometres lie?

CHAPTER XII

Averages

Assume someone has prepared a distribution about which we know nothing—the life-span, say, of a newly discovered type of microbe. We do not want to study the distribution in detail but simply want a few single figures that will give us a good idea of what the distribution is like. What figures would be useful?

Well, first we have no idea as to whereabouts on a time-scale, running from zero to eternity, the microbe life-span distribution lies. A figure giving us some information as to the *location* of the distribution would clearly be useful. Next it would be helpful to know if the distribution was spread out over a large part of the time-scale, or whether in fact the life-spans were clustered closely together. In other words, a single figure summarising the *dispersion* of the distribution would be useful. Finally we may like to know if the majority of the life-spans lie at one end of the distribution or the other or whether they are ranged symmetrically around the middle of the distribution—i.e. we would like a figure indicating the *skew* of the distribution.

In this chapter we will examine measures of location of a distribution and in the next chapter measures of dispersion and skew.

MEDIANS, MEANS AND MODES

1. Measuring location. Finding a single figure that indicates the location of a distribution is not easy. The distribution may be spread out over a wide range of values and to choose one to represent its location is like trying to find a single person who can represent a whole constituency in Parliament. Just as there are different opinions as to which person would make the best constituency representative, so there are different opinions as to which figure best represents the location of the distribution. Such a figure, incidentally, is called an *average*.

In statistics there are three main kinds of averages—each in a different way measuring the location of the distribution. These are the *median*, the *arithmetic mean* (henceforth simply referred to as the "mean") and the *mode.*

2. Median. We have already come across the median (*see* XI, 3). Since the median was defined as the value of the middle item of a distribution it has an obvious claim to be an average.

The median has the following features that make it a particularly useful statistical measure:

(*a*) As we have seen, half the items in a distribution have a value equal to or above the median and half equal to or below the median. (This in turn means that if we choose an item from the distribution at random there would be a 50/50 chance that its value would be equal to or above the median—and vice versa.)

(*b*) It is unaffected by the value of extreme items in the distribution.

NOTE: The problem of extreme values in a distribution often arises in practice. Extreme values frequently occur because of some fault in the collection of the raw data. Thus if in data about ages of infants in a school we find a figure of 54 years it suggests one of the teachers' ages was collected by mistake. Similarly a height of 62 feet in data relating to men's heights suggests a mis-reading of the scale or a clerical error. Now in these cases the figures are obviously erroneous and they will be removed from the data. However, it is often not clear as to whether a figure is an error or genuine, and the statistician is not sure what to do. To include an erroneous figure in the data would distort his results, but to ignore a genuine figure would also cause distortion. Any statistical measure, then, that does not rely on the correctness of extreme items for its value possesses a useful advantage.

3. Finding the median. Finding the median was discussed in XI, 3–5.

4. Mean. The *mean* (or the "arithmetic mean," to give it its full title) is the measure to which we usually refer in everyday life when we use the word "average," and it can be defined as the value each item in the distribution would have if all the values were shared out equally among all the items. Thus, if three people had £2, £3 and £7 respectively the mean amount would be £4—i.e. £12 shared equally between three people. The greatest advantage the mean possesses is that in calculating it every value in the distribution is used and this means that it can be used for further statistical computations (e.g. averaging means—*see* XII, 11).

5. Finding the mean. The arithmetic mean (symbolised as $\bar{x}$) can be found from the formula $\bar{x} = \Sigma x/n$, where "x" stands for the values of the different items in the distribution.

If we use this formula as it stands we are said to be using the "*direct method*" of computing the mean. For example, if we have the following figures: 5, 7, 8, 12, 18, then using the "direct method" we have: $\bar{x} = \Sigma x/n = (5 + 7 + 8 + 12 + 18)/5 = 10$.

In the case of a grouped frequency distribution, this method obviously cannot be used since the values of the individual items are not known. Instead, one of two simple procedures, depending on the form of the distribution, will be used. As these procedures are essentially computational short cuts only, they have been put in Appendix III.

6. Mode. The *mode* is often defined simply as the most frequently occurring value in a distribution. (Note, however, that if the data are continuous it is possible that no two values will be the same. In this sort of situation the mode is defined as the point of maximum frequency density—i.e. where occurrences cluster most closely together.) Being the value that occurs most frequently it is the best representative of the *typical* item. It is this form of average that is implied by such expressions as "the average person" or "the average holiday." The remark that "the average holiday is 3 weeks" means that the *usual* holiday is 3 weeks, not that the mean of all holidays is 3 weeks. The mode is thus a familiar and commonly used average, though its name is less well known.

7. Finding the mode. To find the mode in the case of ungrouped data, simply find the most frequently occurring value. However, finding the mode in the case of a grouped frequency distribution is, unfortunately, not so easy. Since a grouped frequency distribution has no individual values, it is obviously impossible to determine which value occurs most frequently. The best that can be done is to select the *modal class*, i.e. the class with the highest frequency. Yet even this can be unsatisfactory: should a different set of classes have been chosen when the original data was processed, the modal class could appear at a quite different place. Indeed, for practical purposes the student would be well advised to ignore the mode (or place it within wide limits) unless the nature of the data is such that the mode is distinct and definite, as in the case of ungrouped discrete data, for example.

For examination purposes, however, the following method of finding the mode is suggested (*see* Fig. 17):

(*a*) Construct a histogram of the grouped frequency distribution.

(*b*) Find the modal class (i.e. the class with the highest rectangle).

(*c*) Draw a line from the top right-hand corner of the modal class rectangle to the point where the top of the next adjacent rectangle to the left touches it. Draw a corresponding line on the opposite diagonal from the top left-hand corner to the rectangle on the right.

(*d*) The mode is the point where these two lines cross.

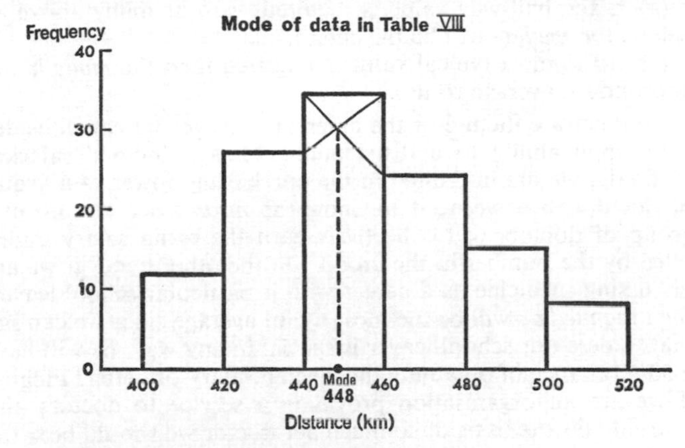

FIG. 17 *Diagrammatic method of finding the mode.* The histogram uses the data in Table VIII.

EXAMPLE

Find the mode of the data in Table VIII.

See Fig. 17, which shows that the mode is 448 kilometres.

If the distribution has a single peak, then the following formula will give the approximate mode:

$$Mode = Mean - 3(Mean - Median)$$

8. Bi-modal distributions. When a frequency polygon (or curve) is superimposed on a histogram there is usually only one peak. Should the distribution be such that two peaks arise, the distribution is said to be *bi-modal.* Distributions normally have a single mode, but a distribution relating to (say) wage rates in a factory that employs both men and women could be bi-modal, one mode relating to the rate paid to the majority of the women and the other to that paid to the majority of the men.

MEAN, MEDIAN AND MODE COMPARED

9. Choice of an average. The above explanations should help us to decide which average should be chosen in any particular case.

(*a*) If we wish to know the result that would follow from an equal distribution—consumption of beer per head, for instance—the *mean* is the most suitable.

(*b*) If the half-way value is required, with as many above as below, the *median* will be the choice.

(*c*) If the most typical value is required then the *mode* is the appropriate average to use.

To illustrate the use of the different averages we can consider their applicability to a distribution of, say, doctors' salaries. If, firstly, we are interested in the purchasing power of a group of doctors then we need to know the *mean* since any normal group of doctors will collectively earn the mean salary multiplied by the number in the group. On the other hand, if we are discussing medicine as a career with a particular school-leaver, then the *median* will be the most useful average since we can say that, unless our school-leaver is special in any way, he will have a 50/50 chance of obtaining the median salary or better. Finally, if we are an organisation providing a service to doctors and charging on the basis of so much per doctor we should bear the *mode* in mind as this is the salary of the typical doctor and the acceptance or rejection of our service may well depend on our charge in relationship to the doctor's salary.

10. Features of the mean, median and mode.

(*a*) *The mean:*

(*i*) It makes use of every value in the distribution. It can, therefore, be distorted by extreme values.

(*ii*) It can be used for further mathematical processing (e.g. *see* **11** below).

(*iii*) It may result in an "impossible" figure where the data is discrete (e.g. 1.737 children).

(*iv*) It is the best known of the averages.

(*b*) *The median:*

(*i*) It is equal to or exceeded by half the values in the distribution—and vice versa.

(*ii*) It uses only one value in the distribution. It is *not*, therefore, influenced by extreme values.

(*iii*) It cannot be used for further mathematical processing.

(*iv*) It is an actual value occurring in the distribution (unless it is computed by averaging the two middle items of an even-numbered distribution).

(*v*) It can be computed even if the data is incomplete. Thus in determining the median salary of a group of executives, for example, it may prove impossible to discover the salaries of the highest-paid executives. But, since these values will not affect the value of the median item, it is still possible to determine the median salary.

(*c*) *The mode:*

(*i*) If the data is discrete then like the median, the mode is an actual, single value.

(*ii*) If the data is continuous then the mode marks the point of the greatest clustering of occurrences.

(*iii*) It can be estimated from incomplete data, but cannot be used for further mathematical processing.

AVERAGING MEANS

11. Procedure for averaging means. If 10 students in one class have an average (arithmetic mean) of 60 marks each and 40 students in a second class have an average of 50 marks each, what is the average mark of the two classes combined?

In this sort of problem there is a strong temptation to compute the mean of the averages, i.e. $\frac{1}{2}(60 + 50) = 55$ marks. This is quite wrong, as can be seen from the following:

(*a*) (*i*) The total mark gained by the whole of the first class must be 60×10, i.e. 600 marks.

NOTE: Since $\bar{x} = \Sigma x/n$, therefore $\Sigma x = \bar{x}n$, where Σx is the total of all the values in the distribution.

(*ii*) Similarly, the total mark gained by the second class must be $50 \times 40 = 2000$ marks.

(*b*) The combined mark of both classes, then, must be 600 + 2000 = 2600 marks.

(*c*) These marks were obtained by a total of 10 + 40 = 50 students, so the mean mark per student must be 2600/50 = 52 marks.

The *correct* procedure for computing such a combined average is, then, as follows:

(*a*) Compute for each group the sum of all the values. This is done by multiplying the group mean by the number of items in the group.

(*b*) Add these totals to give a combined total.

(*c*) Divide the combined total by the combined number of items.

12. Weighted average. This combined mean is often referred to as the *weighted average* because individual group averages are "weighted" by multiplying them by the number of items in the group.

It should be noted that this particular use of a weighted average is really only a specific application of a more general idea. There are other contexts in which, when computing a mean, it is sometimes desired to give certain figures greater importance in the answer (e.g. index numbers). This can be done by multiplying those figures by chosen numbers called *weights.* In the example above, the weights were the number of students in each class, i.e. 10 and 40. The important point to remember about a mean computed from such weights is that, instead of dividing by the number of items, one divides by *the sum of the weights* (unweighted figures carry a "weight" of 1 in this addition). A mean so calculated is called a *weighted average*, and the appropriate formula is:

$$\textit{Weighted average} = \frac{\Sigma xw}{\Sigma w}$$

EXAMPLE

Find the weighted average of the following figures where Group B figures are to carry weights of 2, and Group C figures, weights of 3:

Group A: 6, 5, 3
B: 12, 14
C: 20, 22, 23

Group	*No.* (*x*)	*Weight* (*w*)	*Weight* × *No.* (*xw*)
A	6	1	6
A	5	1	5
A	3	1	3
B	12	2	24
B	14	2	28
C	20	3	60
C	22	3	66
C	23	3	69
		$\Sigma w = 16$	$\Sigma xw = 261$

$$\text{Weighted average} = \frac{\Sigma xw}{\Sigma w} = \frac{261}{16} \simeq 16.3$$

13. Accuracy and the mean. Since the mean is only a representative value of a possibly large number of very different values, it would seem pedantic to state it with extreme accuracy—even though such accuracy was not spurious. For instance, to give the mean height of a group of children as 1.24 m conveys just as much to a reader as 1.2402 m. However, in statistics the mean is frequently used in further calculations (as above, where it was used to obtain a combined mean). For this reason, it is necessary to state it with a good deal more accuracy than would otherwise be called for.

GEOMETRIC AND HARMONIC MEANS

There are two other means apart from the arithmetic mean which, though not used anything like as often as the other averages, are sometimes asked about in examinations. These are the geometric mean and the harmonic mean.

14. Geometric mean (GM).

(*a*) *Computation in theory.*

(*i*) Multiply the values all together (i.e. $x_1 \times x_2 \times x_3 \ldots$) and then,

(*ii*) Find the *n*th root of the product where *n* is the number of items.

(*b*) *Computation in practice.* The theoretical computation clearly involves a great deal of work, but this can be reduced considerably by the use of logarithms. With logs, multiplication is achieved by adding, and the *n*th root found by dividing by *n*. The formula is:

$$\textit{Logarithm of GM} = \frac{\Sigma \log x}{n}$$

To compute the GM, therefore, we must:

(*i*) Add the logs of all the values,
(*ii*) Divide by *n*,
(*iii*) Look up the anti-log of the answer to (*ii*).

EXAMPLE

Find the geometric mean of 4, 5 and 6.

$$\text{Log GM} = \frac{\log 4 + \log 5 + \log 6}{3}$$

$$= \frac{0.6021 + 0.6990 + 0.7782}{3} = \frac{2.0793}{3} = 0.6931$$

$$\therefore \text{GM} = \text{anti-log } 0.6931 = 4.933$$

(*c*) *Use of the GM.* The geometric mean is used mainly in connection with index numbers, though it can also be used for averaging ratios.

15. Harmonic mean.

(*a*) *Computation.*
(*i*) add the reciprocals of the values; then
(*ii*) divide the sum into the number of items.

The formula is:

$$\textit{Harmonic mean} = \frac{n}{\Sigma\frac{1}{x}}$$

EXAMPLE

Find the harmonic mean of 4, 5 and 6.

$$\text{Harmonic mean} = \frac{3}{\frac{1}{4} + \frac{1}{5} + \frac{1}{6}} = \frac{3}{\frac{37}{60}} = 4.86$$

(*b*) *Use of the harmonic mean.* The use of the harmonic mean in statistics is so restricted that discussion is best omitted in a HANDBOOK such as this.

PROGRESS TEST 12

(Answers in Appendix VI)

1. In a series of twenty "spot checks" the following number of passengers were counted at a certain depot:

137	136	135	136
135	135	137	138
136	137	136	136
138	137	136	137
136	136	138	135

(*a*) From these figures determine: (*i*) the mean, (*ii*) the mode, and (*iii*) the median.

(*b*) It is later discovered that the last observation was incorrectly recorded when the data was being collected. It should have been 35 instead of 135. Re-compute the three averages. What features of the three averages do your revised figures bring out?

2. Using the data in Question 1 on p. 75:

(*a*) Compute the mean and median I.Q. of the pupils in each school.

(*b*) If 450 pupils in a third school had a mean I.Q. of 106, what would be the mean I.Q. of all the pupils in the three schools combined?

3. Customers' waiting times in a check-out queue at a Little Fielding self-service store were found to be as follows:

Length of wait (min.)	*No. of customers*
No waiting	50
Waiting under ½	210
½–under 1	340
1– „ 2	200
2– „ 3	110
3– „ 5	170
5– „ 10	140
10 and over	80
	1300

(*a*) A few lucky customers have no waiting and a few unlucky ones have to wait over 10 minutes. How long do the middle 50 per cent of the customers have to wait?

(*b*) (*i*) If you use the store every day, what will your mean waiting time be?

(*ii*) You wish to reach the check-out point by 11.15 a.m. It is now 11.13 as you pick up your last purchase. Are the odds in your favour?

4. A company which makes and sells a standard article has four machines on which this article can be made. Owing to differences in age and design the machines run at different speeds, as follows:

Machine	*Number of minutes required to produce one article*
A	2
B	3
C	5
D	6

(*a*) When all machines are running, what is the total number of articles produced per hour?

(*b*) Over a period of 3 hours, only machines B, C and D were run for the first 2 hours, and only machines A, B and D for the last hour. What was the average number of articles produced *per hour* over this 3-hour period? (*I.C.M.A.*)

5. A distribution of the wages paid to foremen would show that, although a few reach very high levels, most foremen are at the lower levels of the distribution. The same applies, of course, to most wage distributions.

(*a*) If you were an employer resisting a foreman's wage claim, which average would suit your case best?

(*b*) If you were the foremen's union representative, which average would you then select?

(*c*) If you were contemplating a career as a foreman, which average would you examine?

Give reasons for your answers.

6. Show by means of a formula the *combined mean* of three groups, the first group having n_1 items and a mean of $\bar{x}_1$; the second group, n_2 items and a mean of $\bar{x}_2$; and the third group, n_3 items and a mean of $\bar{x}_3$.

CHAPTER XIII

Dispersion and Skew

DISPERSION

NOTE: The word *variation* is sometimes used as an alternative to dispersion.

1. Measuring dispersion. Knowing the average of a distribution in no way tells us whether or not the figures in the distribution are clustered closely together or well spread out.

For example, there could be two groups, each of four men. In the first group, all the men could be 1.676 m high, and in the second group they could have heights of 1.372 m, 1.524 m, 1.829 m and 1.981 m respectively. Both groups have a *mean* height of 1.676 m, but the dispersion in height is much greater in one than the other.

It would clearly be useful to find some way of measuring this dispersion and expressing it as a single figure. Such measures are called *measures of dispersion* (or *variation*) and the most important of these are the:

(*a*) Range.
(*b*) Quartile deviation (semi-interquartile range).
(*c*) Mean deviation.
(*d*) Standard deviation.

2. Range. This is simply the difference between the highest and the lowest values. Therefore:

$$Range = Highest\ value - Lowest\ value$$

EXAMPLE

Find the range of the data subsequently summarised in Table VIII (as Table VIII is a grouped frequency distribution, reference must be made to the original array, Table VI).

Range = 515 − 403 = 112 kilometres.

Unfortunately the range has a grave disadvantage: it is too much influenced by extreme values. If, for example, one single salesman in the Table VIII data had travelled 627 kilometres, the range would have been doubled although the dispersion of the other 119 salesmen would have remained unaltered. Similarly the range can be totally distorted by the sort of errors discussed in the note to XII, **2**.

3. Quartile deviation. This disadvantage of the range can be overcome by ignoring the extreme values. One way of doing this is to cut off the top and bottom quarters by considering only the quartiles, and then see what range is left. This range is called the *interquartile range*. If the interquartile range is divided by 2, the figure obtained is called the *quartile deviation* (or, alternatively, the *semi-interquartile range*), i.e.:

$$\textit{Quartile deviation} = \frac{\textit{Third quartile} - \textit{First quartile}}{2}$$

EXAMPLE

Find the quartile deviation of the figures in Table VIII.

The first and third quartiles of Table VIII were calculated in XI, **8** as 433.3 and 474.1 kilometres respectively.

$$\therefore \text{ Quartile deviation} = \frac{474.1 - 433.3}{2} = 20.4 \text{ kilometres}$$

4. Mean deviation. The *mean deviation* is simply the average (mean) deviation of all the values from the distribution mean. It is found as its description would suggest—by adding up the deviations of all values from the distribution mean and dividing by the number of items. As a formula it can be written as:

$$\text{Mean deviation} = \frac{\Sigma(x - \bar{x})}{n}$$

(Note, however, that the sign of $x - \bar{x}$ must be ignored—i.e. all deviations are written as +.)

EXAMPLE

Find the mean deviation of 5, 7, 8, 12 and 18 (i.e. *the figures used in XII,* **5**).

x	$x - \bar{x}$ (i.e. 10)
5	5
7	3
8	2
12	2
18	8
50	20

$$\bar{x} = \frac{50}{5} = 10$$

$$\therefore \text{ Mean deviation} = \frac{20}{5} = \underline{\underline{4}}$$

If the distribution is a grouped frequency distribution then the procedure given in Appendix III must be employed.

The mean deviation is useful in giving some indication of the extent of the dispersion in terms of all the values in the distribution. It has not, however, any further statistical application and is not, therefore, a very important measure of dispersion.

5. Standard deviation. The *standard deviation* (symbolised as σ) is found by adding the square of the deviations of the individual values from the mean of the distribution, dividing this sum by the number of items in the distribution, and then finding the square root of the quotient. Algebraically this can be shown as:

$$\sigma = \sqrt{\frac{\Sigma(x - \bar{x})^2}{n}}$$

The standard deviation is by far the most important of the measures of dispersion, but its importance is due to its mathematical properties (especially in sampling theory—*see* Part Six) rather than its descriptive properties (i.e. its ability to make a distribution more easily understood). Clearly, the greater the values of individual items differ from the mean, the greater will be the square of these differences and therefore the greater sum of the squares. And the greater this sum, of course, the larger will σ be. Hence the greater the dispersion, the larger the standard deviation will be (note that if there is no dispersion at all—i.e. all the values are the same—then the standard deviation will work out at zero).

6. Computation of the standard deviation. The standard deviation can, of course, be computed directly from the formula given in 5 above. The method is referred to as the "*direct method.*"

EXAMPLE

To illustrate the "direct method" of finding σ we will calculate the standard deviation of the figures previously used in XII, 5, i.e. 5, 7, 8, 12 and 18.

x	$(x - \bar{x})$	$(x - \bar{x})^2$
5	$5 - 10 = -5$	25
7	$7 - 10 = -3$	9
8	$8 - 10 = -2$	4
12	$12 - 10 = +2$	4
18	$18 - 10 = +8$	64
$\Sigma x = 50$		$\Sigma(x - \bar{x})^2 = 106$

$$\therefore \bar{x} = 50/5 = 10$$

$$\sigma = \sqrt{\frac{\Sigma(x - \bar{x})^2}{n}} = \sqrt{\frac{106}{5}} = \sqrt{21.2} = 4.6$$

As in the case of the mean, if the data is in the form of a grouped frequency distribution the "direct method" cannot be used. To compute the standard deviation in these circumstances one of the procedures detailed in Appendix III must be used.

7. Variance. This is the name given to the square of the standard deviation, i.e. Variance $= \sigma^2$. It is important because variances can be added: for instance, if two distributions had variances of σ_1^2 and σ_2^2 respectively, the variance of the two distributions combined would be $\sigma_1^2 + \sigma_2^2$.

8. Units of the measures of dispersion. It should not be forgotten that the measures of dispersion are in the same units as the variable measured. For example, in Table VIII the units are kilometres. The range, quartile deviation and standard deviation are therefore all in kilometres. Had the problem involved a frequency distribution relating to ages of people (measured in years) the measures would all have been in years.

9. Advantages and disadvantages of the different measures of dispersion.

(*a*) *Range.*

(*i*) Very simple to calculate.

(*ii*) Very simple to understand.

(*iii*) Used in practice as a measure of dispersion in connection with statistical quality control. *See* Chapter XXIV.

BUT:

(*iv*) Liable to mislead if unrepresentative extreme values occur.

(*v*) Fails totally to indicate the degree of clustering. (For instance, 3, 5, 5, 5, 5, 7 has the same range as 3, 3, 3, 7, 7, 7, but in the former group values cluster much more closely together.)

(*b*) *Quartile deviation.*

(*i*) Simple to understand.

BUT:

(*ii*) Fails to take into account all the values.

(*iii*) Gives no real indication of the degree of clustering.

(*c*) *Mean deviation.*

(*i*) Indicates the average dispersion of values from the mean.

(*ii*) Uses all the values in the distribution.

BUT:

(*iii*) Cannot be used in further statistical analysis.

(*d*) *Standard deviation.*

(*i*) Is of greatest importance in later statistical work (*see* Part Six).

(*ii*) Uses every value in the distribution.

BUT:

(*iii*) Difficult to comprehend.

(*iv*) Gives more than proportional weight to extreme values because it squares the deviations (e.g. a value twice as far from the mean as another is multiplied by a factor of four—2^2—relative to the latter value).

10. Coefficient of variation. It sometimes happens that we need to compare the variability of two or more sets of figures. For example, are the figures in Table VIII more variable than those given in 6? The standard deviations are respectively 27 kilometres and 4.6 units. But these figures are clearly not comparable since, first, they are in different units and, second, they relate to sets of figures of quite different orders of size.

However, we could obtain some idea of the degree of variability if we could relate the size of a variation to the average of the figures it was derived from, i.e. calculate the standard deviation as a percentage of the mean. This measure is called the *coefficient of variation.* It is expressed by the formula:

$$\textit{Coefficient of variation} = \sigma/\bar{x} \times 100$$

To compare the variability of two sets of figures would therefore involve comparing their respective coefficients of variation.

EXAMPLE

Compare the variability of the figures in Table VIII with those given in **6.**

	Table VIII	*From* **6**
σ	27 km (*See* Appendix II)	4.6
$\bar{x}$	454.5 km („ „ „)	10
Coefficient of variation:		
$\left(\frac{\sigma}{\bar{x}} \times 100\right)$	6	46

Conclusion: The figures used in **6** are very much more variable than those in Table VIII.

SKEW

Finally, we consider the last descriptive measure of a given distribution—its *skew.*

11. Symmetry and skewness. If the histogram of a grouped frequency distribution is drawn, it usually displays quite low frequencies on the left, builds steadily up to a peak and then drops steadily down to low frequencies again on the right. If the peak is in the centre of the histogram and the slopes on either side are virtually equal to each other, the distribution is said to be *symmetrical* (*see* Fig. 18).

On the other hand, if the peak lies to one or other side of the centre of the histogram, the distribution is said to be *skewed* (*see* Figs. 19 and 20). The further the peak lies from the centre of the histogram, the more the distribution is said to be skewed.

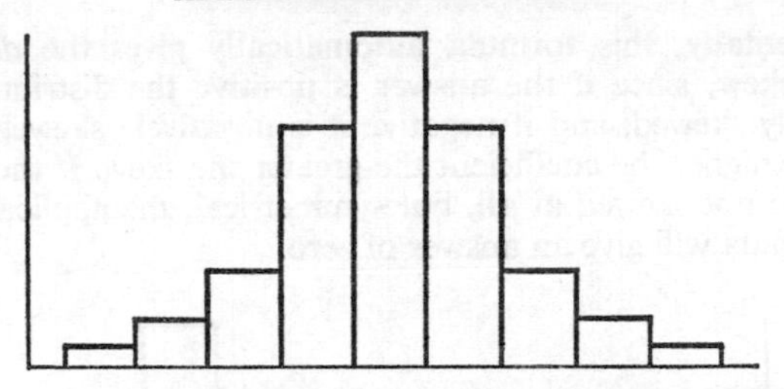

FIG. 18 *Symmetrical distribution.*

The skewness of a distribution can be measured as regards:

(*a*) the *direction* of the skew; and
(*b*) the *degree* of skew.

12. Direction of the skew. The direction of the skew depends upon the relationship of the peak to the centre of the histogram, and is indicated by the terms *positive skew* and *negative skew.*

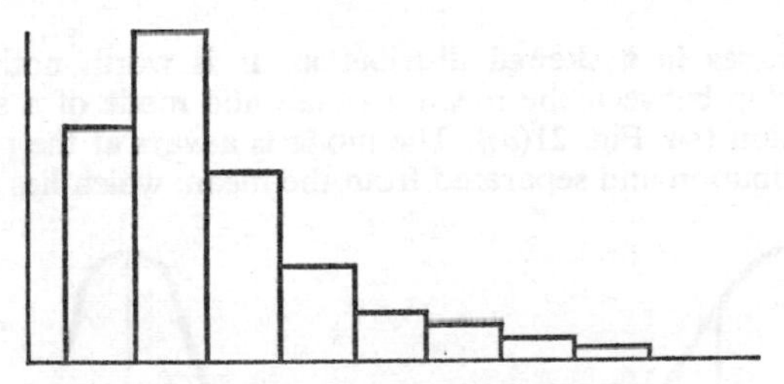

FIG. 19 *Positive skew.*

The skew is *positive* when the peak lies to the left of the centre (*see* Fig. 19) and *negative* when the peak lies to the right of the centre (*see* Fig. 20).

13. Degree of skew. There is more than one way of measuring the degree of skewness, but at this stage the student is advised to learn only one: the *Pearson coefficient of skewness.* It is computed by the formula:

$$Sk = \frac{3\ (Mean - Median)}{Standard\ deviation}$$

where Sk=Pearson coefficient of skewness.

Incidentally, this formula automatically gives the *direction* of the skew, since if the answer is positive the distribution is positively skewed, and if negative it is negatively skewed. Note that the higher the coefficient the greater the skew. If the distribution is not skewed at all, but symmetrical, the application of the formula will give an answer of zero.

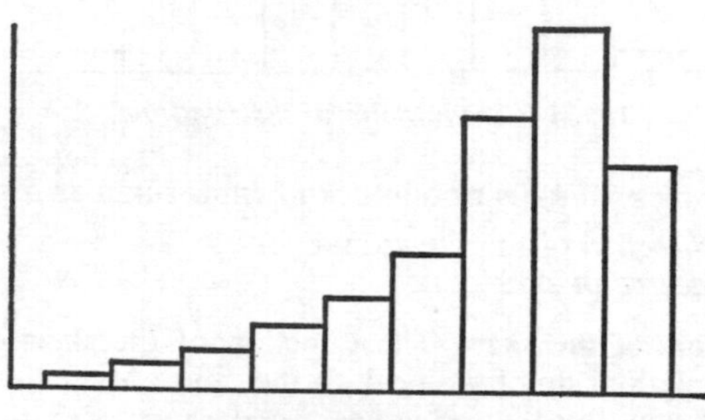

FIG. 20 *Negative skew.*

14. Averages in a skewed distribution. It is worth noting the relationship between the mean, median and mode of a skewed distribution (*see* Fig. 21(*a*)). The mode is always at the peak of the distribution and separated from the mean, which lies on the

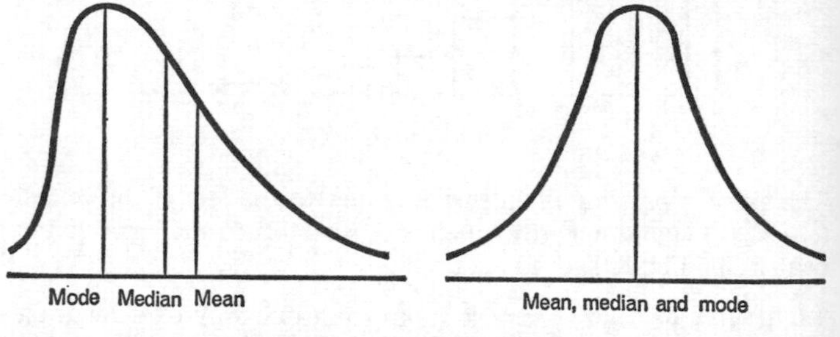

FIG. 21 *Relationship between the mean, median, mode and skew.* (*a*) Skewed distribution. (*b*) Symmetrical distribution.

side of the longer tail, by a distance dependent on the degree of skewness. The median usually lies between the mode and the mean, though it can lie on the mode.

In the case of a symmetrical distribution the student should note that the mean, median and mode all lie at the same point: at the centre of the distribution (*see* Fig. 21(*b*)).

PROGRESS TEST 13

(*Answers in Appendix VI*)

1. Find the standard deviations and the mean deviations of the two distributions in Question 1 on p. 75. Which has the greatest relative variability?

2. Compute the quartile deviation of the following travelling costs:

DAILY TRAVELLING COSTS OF OFFICE STAFF

Pence per day	*Number of staff*
10–under 20	41
20– „ 30	95
30– „ 40	202
40– „ 50	147
50 and over	15

3. Compute the standard deviation of the following data using
(*a*) the "direct method;"
(*b*) an assumed mean:

2450	2461
2460	2449
2455	2452
2441	2451
2448	2455
2440	2443
2452	2458
2444	2463
2446	2453
2459	2440

4. Given that Table VIII has a mean of 454.5 km, a median of 452.4 km and a standard deviation of 27 km, what is the direction and degree of skew of this distribution?

5. Find the coefficient of skewness for the following weekly branch sales:

Sales (£000)	*Branches*
Under 10	25
10–under 20	18
20– „ 30	8
30– „ 40	3
40 and over	1
	55

6. What is the coefficient of skewness of a distribution having a mean of 20, a median of 22 and a standard deviation of 10?

PART FOUR

CORRELATION

CHAPTER XIV

Scattergraphs

Part Three of this book was devoted to finding methods which described a *single* collection of figures. The fundamental idea now in Part Four is the examination of *two* variables—i.e. two collections of figures—and seeing to what extent they are related. Three aspects will be discussed: the scattergraph, regression lines and correlation. In order that the student may compare these aspects the two sets of figures shown in Table X will be used throughout for demonstration.

TABLE X. SALES AND ADVERTISING EXPENDITURE OF THE PQP CO. LTD., 1972–76

Year	*Advertising* (£000)	*Sales* (£000)
	x	y
1972	2	60
1973	5	100
1974	4	70
1975	6	90
1976	3	80

RELATIONSHIP BETWEEN TWO VARIABLES

1. Degrees of relationship. If a car owner were to record daily the petrol he used and the kilometres he covered, he would find a very close relationship between the two sets of figures. As one increased, so would the other. On the other hand, if he compared his daily travelling with, say, the daily number of marriages in New York, he would find there was no relationship at all.

The relationship between kilometres driven and petrol used is an obvious one, but the relationship between other sets of figures is not usually so obvious. Businessmen have found by experience that there is a definite relationship between advertising and sales, but it is often difficult to say how close the relationship is. A study of the figures in Table X will reveal some connection, though an uncertain one, between advertising and sales in the PQP Company. Clearly, some technique is called for to clarify this relationship.

2. Purpose of finding a relationship. Before considering such techniques the student may rightly ask himself, why bother? The answer is that knowledge of the relationship enables us to both *predict* and *control* events.

If we know there is a very close relationship between kilometres travelled and petrol used, it is possible to predict how much petrol will be required for a given journey—or, conversely, how far it is possible to travel using a given quantity of petrol.

Alternatively, knowledge of the relationship can be used to control car performance, since if the distance obtained from a particular petrol consumption subsequently drops below what is expected, it indicates that the engine is not functioning as it should. An overhaul will probably rectify things and so enable the previous performance figures to be attained once more.

3. Prediction involving time lag. In business the ability to predict one figure from another is particularly useful if there is a *time lag* between the two sets of figures.

For instance, there is a close relationship between the number of plans passed by a local authority in one year relating to houses to be built, and the number of baths bought the following year. It is possible for suppliers of baths, therefore, to predict their next year's potential sales from this year's local government statistics (*see also* "Suggested answers," Answer 1 to Progress Test 5, for a further example of time lag).

A similar search goes on in the field of national and international economics for relationships involving time lag, since a knowledge of such relationships enables economists to predict the probable future economic position.

4. Independent and dependent variables. When dealing with two variables it is important to know which is the independent variable and which the dependent.

The *independent* variable is the variable which is *not* affected by changes in the other variable.

The *dependent* variable is the variable which *is* affected by changes in the other.

In the case of Table X, changes in advertising for the year can be expected to affect sales. However a change in the sales will not directly affect advertising expenditure. Thus advertising is the independent variable and sales the dependent.

Sometimes it is not easy to decide which is which. In the case of the car owner above, is the independent variable petrol used or kilometres travelled? In practice, the answer to this sort of question often depends upon the way the data are collected. If predetermined quantities of petrol are put in the tank and the distance travelled is measured, then "petrol used" is the independent variable. If, on the other hand, predetermined distances are driven and the petrol consumption subsequently measured, then the independent variable is "kilometres travelled".

SCATTERGRAPHS

The first technique we shall use to study the relationship between advertising and sales in Table X is the scattergraph. A *scattergraph* is a graph with a scale for each variable and upon which variable values are plotted in pairs.

5. Construction.

(*a*) Construct a graph so that the scale for the independent variable lies along the horizontal axis and the scale of the dependent variable lies on the vertical axis (*see* Fig. 22).

(*b*) Plot each pair of figures as a single point on the graph. Thus, in Table X, in 1972 the £2000 advertising and £60 000 sales form such a pair and therefore a point is plotted on the graph where the £2000 line from the horizontal scale meets the £60 000 line from the vertical scale.

That is really all there is to a scattergraph. Its basic purpose is to enable us to see whether there is *any pattern among the points.* In the case of Table X data, it is clear there is some pattern, as the points tend to rise from left to right (Fig. 22). The more distinct a pattern is, the more closely the two variables are related in some way.

6. The irrelevance of time. There is one point in connection with scattergraph construction it is vital to appreciate, and that is: *time does not enter into the graph at all.* In the case of the data in Table X we are concerned with the relationship between

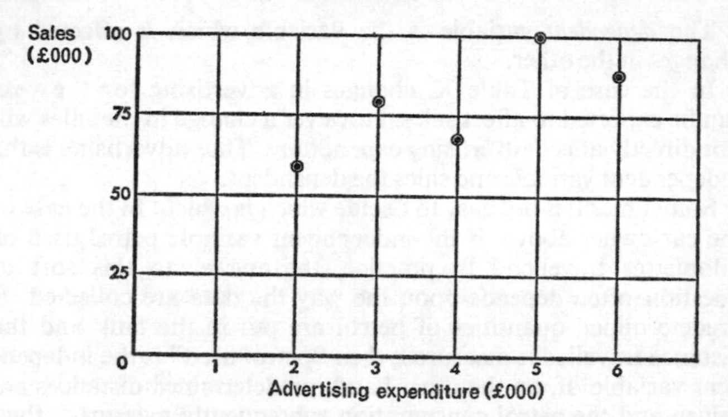

FIG. 22 *Scattergraph of the data in Table X.*

advertising and sales, not *when* these amounts occurred. Thus the points on the scattergraph have no time significance whatsoever. The years in Table X merely linking specific advertising expenditure to specific sales achievements.

7. The line of best fit. The value of a scattergraph can be increased by adding the *line of best fit*. This is the line judged to fit best the pattern of the points; it is drawn so as to pass centrally through the graph of points (*see* Fig. 23). Since all the points cannot lie on the line, the object is to minimise the total divergence of the points from the line.

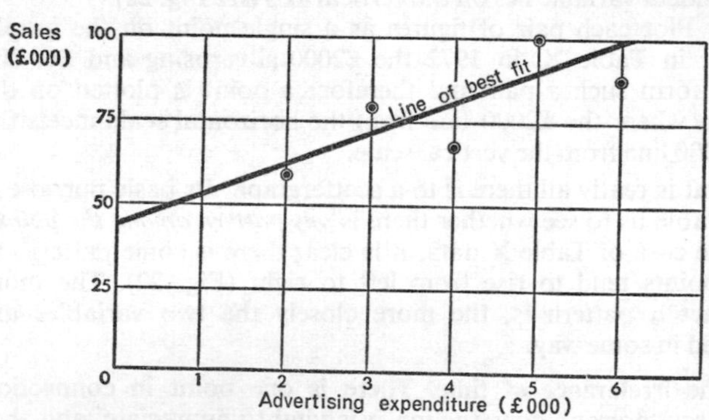

FIG. 23 *Scattergraph of the data in Table X, with line of best fit added.*

8. Estimates from a scattergraph. Once the line of best fit has been inserted, our scattergraph can be used for estimating simply by reading off from the line the sales value corresponding to any level of expenditure on advertising. Supposing £4500 were to be spent on advertising, it can be seen from the line of best fit that this value is associated with sales of £85 000. Therefore £85 000 is the estimated sales to be obtained from an advertising expenditure of £4500.

NOTE: Earlier the usefulness of prediction was discussed. A prediction is, of course, an estimate relating to future events. However, since the techniques we shall be discussing involve estimates relating to past, present and future events, the word "prediction" will give way in the text to the more general word "estimate."

9. The reliability of estimates. The student may not be too happy with this estimate. He may well have noticed from the scattergraph that sales nearly as large were once associated with advertising of only £3000; whereas on another occasion, advertising expenditure of as much as £6000 brought in only £90 000 worth of sales. How reliable, he may ask, is the estimate of £85 000?

The answer is: "Not very." The relationship here between advertising and sales is just not close enough for estimates to be made with great accuracy (as, for example, the estimates discussed earlier relating to petrol and kilometres would be). Yet it is still true that the *best* estimate we can make about sales, given £4500 of advertising, is that they will be £85 000, i.e. any other figure estimated for sales would be even more subject to error than the £85 000.

10. Significance of the "scatter." The next question is, can the reliability of a prediction be gauged? The answer is that to some extent it can be gauged from the "scatter" of the points. When the points on a scattergraph are so dispersed that the pattern is only vaguely discernible, the two variables are *not* very closely related and the line of best fit is an unreliable guide for making estimates.

On the other hand, when the two variables are closely related the points will lie virtually on the line of best fit, and estimates made from such a graph will be very reliable.

11. Curvilinear relationship. There are occasions when the line of best fit is a curve rather than a straight line (*see* Fig. 24).

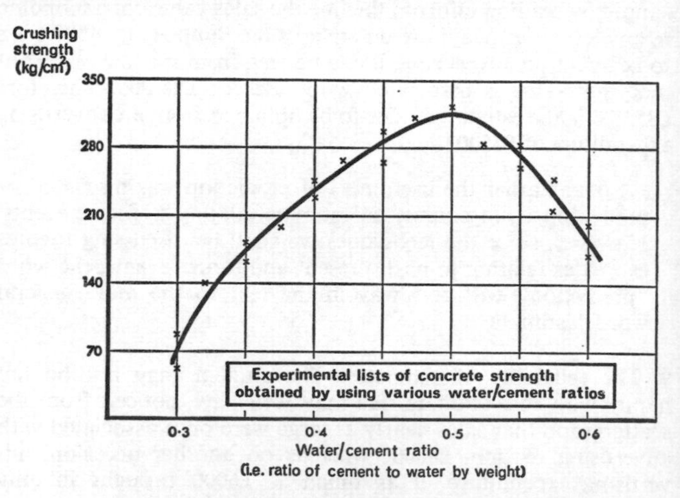

FIG. 24 *Curvilinear relationship.*

When the points on a scattergraph lie close to such a curve, estimates can be quite reliable. However, this type of relationship (termed *curvilinear*) lies beyond the scope of this book and so will not be discussed further. All the relationships we shall consider will be straight-line, or *linear*, relationships.

12. Limitations of scattergraphs. A thoughtful student may by now have realised that scattergraphs have two serious limitations, namely:

(*a*) *Uncertainty as to the correct position of the line of best fit.* If the best estimates are to be made from the line of best fit, it must obviously first be drawn in the correct place. So far this has depended on the judgment of the person constructing the graph. It would be much better if a mathematical method of drawing the line could be devised instead of leaving it to the artistic whim of the individual.

(*b*) *Lack of a measure of the closeness of the relationship.* Since the reliability of an estimate depends heavily on the closeness of the relationship between variables, the lack of any measure of the closeness limits the value of scattergraphs considerably.

In the next two chapters we see how these limitations can be overcome.

PROGRESS TEST 14

(Answers in Appendix VI)

1. The I.Q.s of a group of six people were measured, and they then sat a certain examination. Their I.Q.s and examination marks were as follows:

Person	*I.Q.*	*Exam marks*
A	110	70
B	100	60
C	140	80
D	120	60
E	80	10
F	90	20

Construct a scattergraph of this data, and draw the line of best fit.

(*a*) What marks do you estimate a candidate with an I.Q. of 130 would obtain?

(*b*) Estimate the I.Q. of a candidate who obtained a mark of 77.

2. Two dice were thrown and the sum of their pips doubled to give "dice value." At the same time a card was drawn at random out of a normal pack from which all court cards had been removed. The value of this card was called the "card value."

Throw	*Dice value*	*Card value*
1st	8	8
2nd	8	9
3rd	14	10
4th	22	5
5th	22	8
6th	16	3
7th	12	3
8th	6	2
9th	10	7
10th	10	5

Construct a scattergraph of this data. Is it possible to draw a line of best fit? What conclusions can you draw about the two variables?

CHAPTER XV

Regression Lines

NOTE: It is not necessary to study this chapter to understand the rest of the book. Students who are not required to study regression lines may turn at once to the next chapter.

At the end of the previous chapter it was pointed out that a line of best fit drawn by eye is dependent upon the subjective judgment of the person who draws it. Consequently the position of the line will differ slightly from person to person. A line of best fit independent of individual judgment will have to be drawn mathematically. Such a line is called a *regression line.*

COMPUTING REGRESSION LINES

1. Equation of the line. The general equation for any straight line on a graph is

$$y = a + bx$$

where a and b are constants. Describing the line of best fit means, therefore, finding the appropriate values of a and b.

NOTE: It is assumed that students know this equation. To revise briefly, the a and the b are similar to the components of a two-part electrical tariff where there is a fixed charge per quarter of £a and a further charge of £b per unit of electricity used. If x is the number of units used in a quarter, then y (the total charge) $=$ £a $+$ £$b \times x$, or $y = a + bx$. All straight lines on graphs can be expressed in this way, using appropriate values for a and b. So, to describe any line, it is only necessary to find the a and b associated with that line.

2. The method of least squares. When drawing a line of best fit, an attempt is made to minimise the total divergence of the points from the line (*see* XIV, 7). In computing the line mathematically, the same idea is pursued, only it has been found that the best line is one that minimises the total of the *squared* deviations. This computation is logically known as the *method of least squares*.

3. Measuring the deviations. When it comes to measuring the deviations of the points from the line, it is important to understand that statisticians do *not* measure the shortest distance between a point and the line of best fit. What they measure is either

(*a*) the *vertical* distance between point and line, or
(*b*) the *horizontal* distance (*see* Fig. 25).

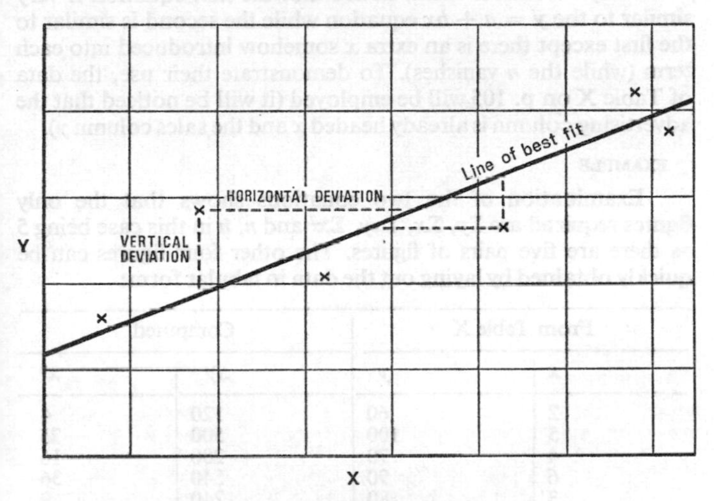

FIG. 25 *Measuring horizontal and vertical deviations from the line of best fit.*

These two different ways of measuring the deviations will produce *two different regression lines*—one minimising the total of the squared deviations measured vertically; the other minimising the total of the squared deviations measured horizontally.

This can be rather confusing, so for the moment only the calculations needed to determine regression lines will be considered. The distinction between them will be dealt with in 7 *et seq*.

First, let us take the regression line in which deviations are measured vertically. This is by far the commoner of the two.

4. The regression of y on x. In order to find a and b in the regression line equation it is necessary to solve two simultaneous equations. They are:

$$\Sigma y = an + b\Sigma x$$
$$\Sigma xy = a\Sigma x + b\Sigma x^2$$

where $n =$ the number of *pairs* of figures.

Although these may appear formidable both to remember and use, they look worse than they really are (*see* NOTE, p. 115).

To help remember them, notice that the first equation is very similar to the $y = a + bx$ equation while the second is similar to the first except there is an extra x somehow introduced into each term (while the n vanishes). To demonstrate their use, the data of Table X on p. 105 will be employed (it will be noticed that the advertising column is already headed x and the sales column y).

EXAMPLE

Examination of the two equations shows that the only figures required are Σy, Σx, Σxy, Σx^2 and n, n in this case being 5 as there are five pairs of figures. The other four figures can be quickly obtained by laying out the data in tabular form:

From Table X		Computed	
x	y	xy	x^2
2	60	120	4
5	100	500	25
4	70	280	16
6	90	540	36
3	80	240	9
$\Sigma x = 20$	$\Sigma y = 400$	$\Sigma xy = 1680$	$\Sigma x^2 = 90$

It only remains now to insert these values into the two equations and solve for a and b:

(*i*) $400 = a \times 5 + b \times 20$
(*ii*) $1680 = a \times 20 + b \times 90$

To solve:

Multiply equation (*i*) by 4, i.e. $1600 = 20a + 80b$
Multiply equation (*ii*) by 1, i.e. $1680 = 20a + 90b$
Subtracting upper from lower: $80 = 10b$
$\therefore b = 8$

Now substitute 8 for b in equation (i), i.e.

$$400 = 5a + 8 \times 20$$
$$\therefore \quad 5a = 400 - 160 = 240$$
$$\therefore \quad a = 48$$

The regression line, therefore, is $y = a + bx = 48 + 8x$.

This line is known as the *regression line of y on x* and is the line of best fit when the deviations are measured vertically.

NOTE: Students who doubt their ability (i) to solve simultaneous equations can derive a and b from these two formulae:

$$a = \frac{\Sigma y - b \times \Sigma x}{n}$$

$$b = \frac{n \times \Sigma xy - \Sigma x \times \Sigma y}{n \times \Sigma x^2 - (\Sigma x)^2}$$

(ii) to be accurate when large numbers are involved can use Method 6 in Appendix III.

5. The regression of x on y. If it is desired to compute the second regression line, known as the *regression line of x on y*, it is only necessary to alter the two simultaneous equations so that the x's and y's are interchanged. The equations therefore become:

$$\Sigma x = an + b\Sigma y$$
$$\Sigma xy = a\Sigma y + b\Sigma y^2$$

EXAMPLE

Taking the data in Table X again, the only new figure required is Σy^2. This comes to 33 000, as the student can easily check. Therefore the equations are:

(i) $20 = a \times 5 + b \times 400$
(ii) $1680 = a \times 400 + b \times 33\,000$

To solve:

Multiply equation (i) by 80, i.e. $1600 = 400a + 32\,000b$
Multiply equation (ii) by 1, i.e. $1680 = 400a + 33\,000b$
Subtracting upper from lower: $80 = 1000b$
$\therefore b = 0.08$

Substituting 0.08 for b in equation (i):

$$20 = 5a + 0.08 \times 400$$
$$\therefore \quad 20 = 5a + 32$$
$$\therefore \quad a = -2.4$$

Therefore the equation of the line is $x = -2.4 + 0.08y$.

This equation is *not* merely the previous one turned round. It is quite different, as the student can prove for himself if he wishes.

6. Graphing regression lines. It is quite easy to graph the regression lines once they have been computed. All one has to do is:

(*a*) choose any two values (preferably well apart) for the unknown variables on the right-hand side of the equation,
(*b*) compute the other variable,
(*c*) plot the two pairs of values, and
(*d*) draw a straight line through the plotted points.

EXAMPLE

Graph of regression lines for Table X as computed above:

(*a*) *Regression line of y on x*

(*i*) Let $x = £6000$
$\therefore\ y = 48 + 8 \times 6$ (remember, x was in '000s)
$= £96\,000$ (y too was in '000s)
(*ii*) Let $x = £0$
$\therefore\ y = 48 + 0$
$= £48\,000$

These points, and the regression line through them, are shown in Fig. 26.

(*b*) *Regression line of x on y*

(*i*) Let $y = £100\,000$
$\therefore\ x = -2.4 + 0.08 \times 100$
$= £5600$
(*ii*) Let $y = £50\,000$
$\therefore\ x = -2.4 + 0.08 \times 50$
$= £1600$

Again, these points, and the regression line through them, are shown in Fig. 26.

THE USE OF REGRESSION LINES

7. Use of the regression line of *y* on *x*. Use of the regression line first calculated in **4** is quite simple. To revert to our advertising figures (Table X), it is only necessary to replace x in the equation by a value for advertising to obtain an estimate of the sales, y. If we again take an advertising figure of £4500, the estimated sales will be:

$$y = 48 + 8 \times 4\tfrac{1}{2}$$
$$= 48 + 36 = 84, \text{ i.e. sales of £84 000.}$$

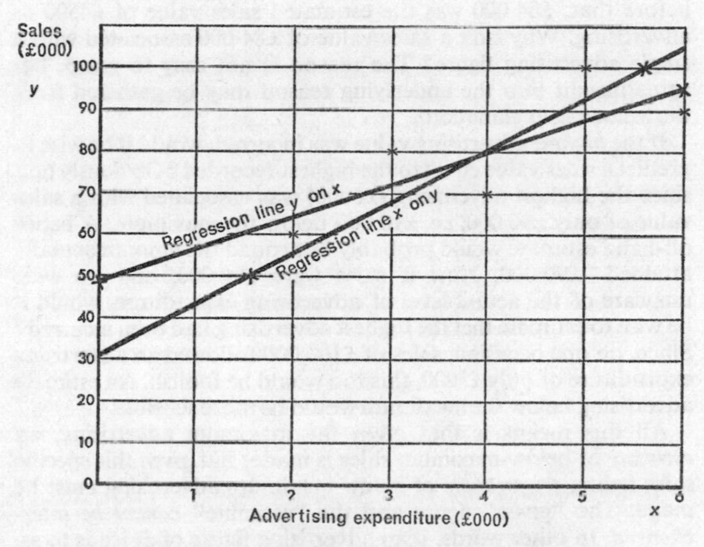

FIG. 26 *Regression lines of the data in Table X.*

NOTE as a point of interest that regression lines always intersect at the *means of the series*—here, £4000 and £80 000.

8. Use of the regression line of x on y. In a similar way the regression line of x on y is used to estimate the value of x that follows from any given value of y. In the example so far discussed, it means estimating the advertising that would be associated with any given sales value. If, for example, sales had been £84 000, the advertising estimated to have been incurred would be:

$$x = -2.4 + 0.08 \times 84$$
$$= 4.32, \text{ i.e. £4320}$$

It will be seen that this is a different figure from the £4500 advertising that gave an estimated £84 000 sales. This paradox is one of the reasons why regression lines seem so confusing to students.

9. The regression line paradox. In the previous paragraph an advertising expenditure of £4320 was estimated to have been incurred, given a sales value of £84 000. Yet in the paragraph before that, £84 000 was the estimated sales value of £4500 of advertising. Why isn't a sales value of £84 000 associated with a single advertising figure? The reason is not easy to grasp, but some insight into the underlying reason may be gathered from the following explanation.

If the *highest* advertising value was incurred, would it be wise to predict a sales value equal to the highest recorded? Obviously not, since the highest advertising (£6000) was associated with a sales value of only £90 000, i.e. £10 000 below the maximum. A better off-hand estimate would probably be around the amount actually attained—£90 000. Now if sales were £90 000, and we were unaware of the actual level of advertising expenditure, would it be wise to estimate that the highest advertising had been incurred? Since, on one occasion, sales of £100 000 followed an advertising expenditure of only £5000, this too would be foolish. An estimate advertising below the maximum would be more sensible.

All this means is that *given* the maximum advertising, an *estimate* of below-maximum sales is made; but *given* this specific sales figure, an *estimate* of below-maximum advertising must be made. The "given" figure and the "estimate" *cannot be interchanged.* In other words, if an advertising figure of £i leads to an estimate of £j sales, you cannot, on learning that in a certain year sales of £j were made, estimate advertising to have been £i.

Regression lines are like one-way streets: you can only move in the authorised direction. If you want to reverse direction you must move across to the other regression line.

10. Choice of regression line. We have seen that there are two regression lines and that it is important to choose the right one. The next problem is *how* to choose the right one. Luckily the answer is quite simple: always use the line that has *the variable to be estimated on the left-hand side of the equation.* If the y variable is to be estimated, use the $y = a + bx$ ("y on x") line; if x is to be estimated, use the $x = a + by$ ("x on y") line.

Indeed, remembering which line is which can perhaps be made easier still by noting the following left-hand/right-hand patterns of regression line facts:

Left-hand	*Right-hand*
Estimate y	given x:

Use $y = a + bx$ equation,

which gives y on x regression line and

$$\text{is found from} \begin{cases} \Sigma y = an + b\Sigma x \\ \Sigma xy = a\Sigma x + b\Sigma x^2 \end{cases}$$

Left-hand	*Right-hand*
Estimate x	given y:

Use $x = a + by$ equation,

which gives x on y regression line and

$$\text{is found from} \begin{cases} \Sigma x = an + b\Sigma y \\ \Sigma xy = a\Sigma y + b\Sigma y^2 \end{cases}$$

11. Reliability of estimates. Although we have just been through a quite lengthy procedure for obtaining estimated values, it is important to realise that such estimates are no more reliable than those derived from a well constructed scattergraph. The procedure simply ensures that the scattergraph *is* well constructed. The reliability of estimates depends far more on the closeness of the relationship of the figures between the variables than on the most elaborate mathematical calculations. The only reason for computing a regression line is to obtain a line of best fit free of subjective judgment (it must be admitted, though, that the regression line does improve the estimate to the extent that it obviates the possibility of a badly judged line of best fit).

A more important reason, though it does not concern us here, is that in advanced statistics the reliability of estimates made from such a line can be measured mathematically.

12. Regression coefficient. The term regression coefficient occasionally arises in statistical discussion. This is simply the value of b in the equations just discussed. For example, in the case of the regression of sales on advertising, the regression coefficient was 8 and in the case of the regression of advertising on sales it was 0.08.

PROGRESS TEST 15

(Answers in Appendix VI)

1. Determine the regression lines relating to the data in Question 1 on p. 111 and estimate (*a*) the marks that would be obtained by a candidate with an I.Q. of 130, (*b*) the I.Q. of a candidate who obtained 77 marks.

Superimpose these lines on the scattergraph constructed for Question 1 on p. 111.

2. Using the data in Question 2 on p. 111.

(*a*) Compute the regression line of card value on die value and use this line to determine the best estimate of the card value that would be associated with a die value of (*i*) 4, (*ii*) 24.

(*b*) Compute the regression line of dice value on card value and use the line to determine the best estimate of the die value that would be associated with a card value of (*i*) 1; (*ii*) 10.

(*c*) Using this example as an illustration explain why it is necessary to compute two regression lines—one for estimating a y value from an x value, and the other an x value from a y value?

Superimpose these regression lines on the scattergraph constructed for Question 2 on p. 111.

CHAPTER XVI

Correlation

We now consider the second limitation of scattergraphs mentioned at the end of Chapter XIV. There it was pointed out that the reliability of an estimate depended on the closeness of the relationship between two variables and that a measure of this closeness was essential for assessing reliability. The *coefficient of correlation* is such a measure. There are different measures of correlation, but the most generally used is one called the *Pearson product-moment coefficient of correlation*, commonly symbolised as r.

COMPUTATION OF r

1. Formula. The formula for calculating r can be expressed in a number of ways. The most practical is:

$$r = \frac{\Sigma xy - n\bar{x}\bar{y}}{n\sigma_x\sigma_y}$$

where $n =$ the number of pairs.

2. Procedure. Examination of the formula indicates that to find r the following steps should be taken:

(*a*) Calculate the means of each series of figures ($\bar{x}$ and $\bar{y}$).

(*b*) Calculate the standard deviations of each series (σ_x and σ_y).

(*c*) Calculate Σxy; in other words multiply each number in the first series with its counterpart in the other (i.e. multiply pairs together) and then add the products.

(*d*) Apply the formula.

EXAMPLE

Find r *of the data in Table X on p. 105.*

(*a*) Find the means.

x (advertising)	*y (sales)*
2	60
5	100
4	70
6	90
3	80
$\Sigma x = 20$	$\Sigma y = 400$

$$\therefore \bar{x} = \frac{20}{5} = \underline{\underline{4}} \qquad \therefore \bar{y} = \frac{400}{5} = \underline{\underline{80}}$$

(*b*) Find the standard deviations. Since the means here are round numbers, the easiest method of calculating the standard deviations is by the "direct method" (*see* XIII, 6):

x	$x - \bar{x}$	$(x - \bar{x})^2$	y	$y - \bar{y}$	$(y - \bar{y})^2$
2	−2	4	60	−20	400
5	+1	1	100	+20	400
4	0	0	70	−10	100
6	+2	4	90	+10	100
3	−1	1	80	0	0

$$\Sigma(x - \bar{x})^2 = \underline{10} \qquad \Sigma(y - \bar{y})^2 = \underline{1000}$$

$$\therefore \sigma_x = \sqrt{\frac{10}{5}} = \underline{\underline{\sqrt{2}}} \qquad \therefore \sigma_y = \sqrt{\frac{1000}{5}} = \underline{\underline{\sqrt{200}}}$$

(*c*) Find Σxy.

x	y	xy
2	60	120
5	100	500
4	70	280
6	90	540
3	80	240

$$\Sigma xy = \underline{\underline{1680}}$$

(*d*) Apply formula $r = \dfrac{\Sigma xy - n\bar{x}\bar{y}}{n\sigma_x\sigma_y}$:

$$\therefore r = \frac{1680 - 5 \times 4 \times 80}{5 \times \sqrt{2} \times \sqrt{200}} = \frac{1680 - 1600}{5 \times \sqrt{400}} = \underline{\underline{+0.8}}$$

NOTE: *r* is not expressed in any units.

The computation of *r* is rather a lengthy business, but provided the formula is known it is not particularly complicated. It is simply a matter of computing three distinct sets of figures and then using them in the formula.

INTERPRETATION OF *r*

3. Need for experience. It has been repeatedly emphasised that the reliability of estimates depends upon the closeness of the relationship between two sets of figures; now that we have a measurement of this closeness, the student is probably keen to learn how to interpret this figure—particularly in assessing the reliability of estimates.

Unfortunately, its interpretation depends very much on experience. The full significance of *r* will be grasped only after working on a number of correlation problems and seeing the kinds of data which give rise to various values of *r*. Until this experience has been gained the student would be wise to interpret *r* very cautiously and to restrict such interpretation to the most general terms. However, to give him a little insight into the significance of *r*, its interpretation will now be discussed in just such general terms.

4. Phrases to describe correlation. Before we examine the numerical significance of *r* it would be advisable to define certain phrases that are commonly used to describe correlation. Even better than verbal description are actual examples shown on a scattergraph, and so some of the following terms of correlation are therefore illustrated in Fig. 27.

(*a*) *Positive correlation.*—When an increase in one variable is associated to a greater of lesser extent with an increase in the other (e.g. advertising and sales).

(*b*) *Negative correlation.*—When an increase in one variable is associated to a greater or lesser extent with a decrease in the other (e.g. TV registrations and cinema attendance).

(*c*) *Perfect correlation.*—When a change in one variable is matched by a change of equal degree in the other variable. If both increase or decrease together, it is *perfect positive correlation*; if one decreases as the other increases, it is *perfect negative correlation*. Perfect correlation only exists when all the points of a scattergraph lie on the line of best fit.

(*d*) *High correlation.*—When a change in one variable is usually associated with a change of a similar, but not equal, degree in the other (e.g. petrol used and miles travelled).

(*e*) *Low correlation.*—When a change in one variable is rarely associated with a change of similar degree in the other (e.g. holiday makers and marriage ceremonies).

(*f*) *Zero correlation.*—Here the variables are not correlated at all, and there is no relationship between changes in one variable and changes in the other (e.g. London bus fares and New York rainfall).

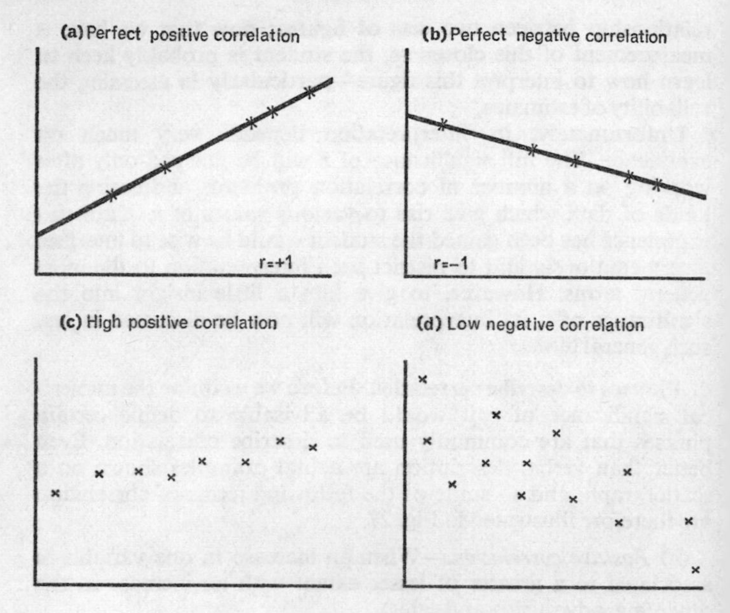

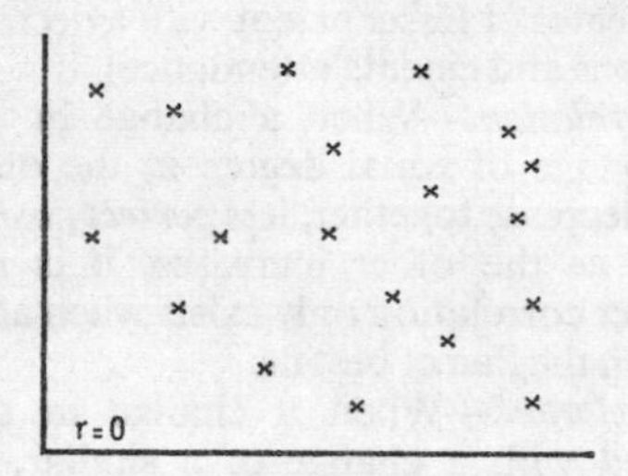

FIG. 27 *Types of correlation.*

5. Numerical values of *r*. The formula for *r* is such that its value always lies between −1 and +1.

+1 means there is *perfect positive* correlation.
−1 means there is *perfect negative* correlation.
0 means there is *zero* correlation.

The closer r is to $+1$ or -1, the closer the relationship between the variables; and the closer r is to 0, the less close the relationship. Beyond this it is not safe to go. The full interpretation of r depends on circumstances (one of which is the size of the sample: *see* Question 4 in Progress Test 16), and all that can really be said is that when estimating the value of one variable from the value of another, then the higher r is the better the estimate will be.

NOTE: r is not really as vague and elusive as would appear at this stage. In rather more advanced statistics, it is possible to state exactly how reliable an estimate is.

One final point should be made. The closeness of the relationship is not proportional to r: an r of (say) 0.8 does *not* indicate a relationship twice as close as one of 0.4 (it is, in fact, very much closer).

6. Spurious correlation. When interpreting r it is vitally important to realise that there *may be no direct connection at all between highly correlated variables.* When this is so the correlation is termed *spurious* or *nonsense* correlation. It can arise in two ways:

(*a*) There may be an *indirect connection.* For example, motor way driving and holidays on the Continent are probably quite highly correlated, since both tend to increase with a rising standard of living. To draw the conclusion that motorway driving *causes* Continental holidays or vice versa would be quite wrong, and a decision (say) by the Chancellor of the Exchequer to close Britain's motorways as part of a campaign to reduce British expenditure abroad would be ridiculous.

(*b*) There may be a *series of coincidences.* Normally, spurious correlation cannot arise in the course of properly conducted statistical work. Laymen, however, sometimes fall into the following trap. They examine the series relating to the variable they wish to predict—say the weather in June—and then hunt about amidst all sorts of data looking for a series that correlates highly with this variable. Needless to say, it is quite likely that one set of data out of the hundreds at hand will fit, even if it is something as unlikely as the number of divorces in January. Then they draw everyone's attention to the high correlation and proceed to predict June weather on the basis of it. Such predictions are, of course, quite unreliable.

7. Correlation and regression lines. Those who read Chapter XV will probably be interested to know that the higher the correlation, the smaller the angle between the two regression lines. In the case of perfect correlation (positive or negative) the two lines will coincide and there will be only one line.

Conversely, the lower the correlation, the greater the angle between the two lines—until, when r is zero, the lines lie at right angles to each other. The "y on x" line will be horizontal and the "x on y" line vertical.

RANK CORRELATION (r')

8. Ranking. In some sets of data the actual values are not given, only the order in which the items are ranked.

EXAMPLE

The examination results in Physics and French of a class of seven boys could be given as places, e.g.

Boy	*Physics*	*French*
Allen	3	1
Birch	2	4
Clark	1	2
Davis	4	3
Evans	6	5
Ford	5	7
Gregory	7	6

9. Computation of rank correlation. Although r could be computed by using the usual formula and treating such rankings as the x and y series of figures, a quicker way is to use the following formula:

$$r' = 1 - \frac{6\,\Sigma d^2}{n(n^2 - 1)}$$

where r' = the coefficient of rank correlation and
d = the *difference* between the rankings of the same item in each series.

The computation of r' in the above example is, therefore:

Physics rank	*French rank*	*d*	*d²*
3	1	2	4
2	4	2	4
1	2	1	1
4	3	1	1
6	5	1	1
5	7	2	4
7	6	1	1
			$\Sigma d^2 = 16$

Applying formula $r' = 1 - \frac{6\,\Sigma d^2}{n(n^2 - 1)}$, then

$$r' = 1 - \frac{6 \times 16}{7(7^2 - 1)} = 0{\cdot}714$$

r' is, of course, interpreted in just the same way as r.

10. Notes on r'.

(*a*) Since there is some loss of information when rankings only are given, r' is not as accurate as the figure that would be obtained by calculating r from the full sets of *marks* in both examinations.

(*b*) Taking the rankings just given, the student should compute the coefficient of correlation between them, using the usual r formula, and prove to his own satisfaction that the answer is the same as above.

PROGRESS TEST 16

(*Answers in Appendix VI*)

1. Compute the coefficient of correlation for the data in Question 1 on p. 111.
2. Compute the coefficient of correlation for the data in Question 2 on p. 111 and state what type of correlation exists between the two sets of figures.
3. Lay out the data in Question 1 on p. 111 in the form of rankings and compute the rank correlation. Why does the correlation figure obtained not equal the figure obtained in the answer to Question 1?
4. Determine the coefficient of correlation for the data in the following table:

Year	*x*	*y*
1975	5	17
1976	75	19

Explain why your answer comes to the figure it does. What conclusion can you draw from this exercise regarding the interpretation of r in small samples?

5. In the course of a survey relating to examination success, you have discovered a high *negative* correlation between students' hours of study and their examination marks. This is so at variance with common sense that it has been suggested an error has been made. Do you agree?

PART FIVE

PROBABILITY

CHAPTER XVII

Basic Probability Theory

In this Part the theory of probability is very briefly discussed. As was explained in the Preface, it is not possible to provide in this book more than a sketch of the theory. For a more detailed exposition the student is referred to the author's book *Operational Research* in this series.

NOTE: Knowledge of this Part is not necessary for understanding virtually the whole of the rest of the book.

PROBABILITY DEFINITIONS

It is very likely that the secret of grasping the undoubtedly complex subject of probability lies in fully understanding the meanings of the various terms used. In this section the more fundamental terms are defined.

1. What is probability? It seems impossible to frame an adequate definition of *probability*. It is a word that appears to be intuitively understandable rather than rigorously definable. However, the concept is constantly being used by people without much trouble (note the easily understandable words "likely" "undoubtedly" and "impossible"—all used in the last few lines) and so we will make no attempt to define probability here.

If not easily definable, probability can be measured and the most useful measure involves relating how often an event *will normally occur* to how often it *could* occur—and expressing this relationship as a fraction. Thus, if we toss a coin 1000 times then normally we will obtain about 500 heads. So the probability of tossing a head is measured as $\frac{500}{1000} = \frac{1}{2}$. Again, in the case of a pack of cards, we will normally draw the ace of spades once in 52 draws, so the probability of drawing the ace of spades is $\frac{1}{52}$.

NOTE: In using this measurement of probability no event can have a probability of less than 0 (since an event cannot occur less often than never) nor more than 1 (since an event cannot occur more often than every time). So all probabilities will lie between 0 (impossible) and 1 (certain).

2. Events. Accurate identification of the type of events involved in a probability problem is essential. The following definitions are, therefore, very important and should be thoroughly learnt:

(*a*) *Event*. An *event* is simply an occurrence. Tossing a head, then, is an event as is throwing a 3 and a 4 with two dice.

NOTE: A failure is also an event—e.g. not tossing a head is an event.

(*b*) *Elementary event*. An *elementary event* is an event involving a single element. Tossing a head is an elementary event.

(*c*) *String-event*. A *string-event* is an event involving a *string of elementary events*. Throwing a 3 and a 4 with two dice is a string event since it is composed of the elementary event "throwing a 3 with one die" and the elementary event "throwing a 4 with the other die." Two points should be noted about string-events:

(*i*) The order of the elementary events is an important feature of a string-event. For example, throwing a 3 with the first die and then throwing a 4 with the second is *not* the same as throwing a 4 with the first die and then throwing a 3 with the second. Each is a different string-event. Indeed, it will probably pay the student to envisage a string-event as elementary events threaded literally on a string—so that a different order of elementary events gives a different string-event.

(*ii*) The concept of string-events has been specifically devised by the author for explanatory purposes. This addition to probability terminology has been made because a group of elementary events in a specified order can be *collectively* treated as if it were itself a single event. To simply call it an "event," however, could lead to confusion with the elementary events that comprised it. It has been thought preferable to give the group a name of its own and "string-event" seemed a reasonable choice. The student is warned, however, that references to string-events are not made in other literature.

(*d*) *Exclusive events*. *Exclusive events* are events that can *never* occur together. Thus a 3 and a 4 with a single throw of a die are exclusive events since both cannot occur together. On the other hand throwing a 3 with one die and a 4 with another are not exclusive events. Note that in a two-dice situation the string-events first-3-then-4 and first-4-then-3 are exclusive events since one throw of the two dice cannot give *both* string-events simultaneously.

(*e*) *Inclusive events. Inclusive events* are events that can occur together, such as throwing a 3 with one die and a 4 with another. It should be appreciated that it is irrelevant whether simultaneous occurrences are likely or not—the fact that they *can* occur together makes them inclusive. Thus, it is unlikely that I can run a four-minute mile and be fast asleep at the same time but since it could just happen these two events are inclusive.

3. *p* and *q*—success and failure. In probability theory if an event with which we are concerned occurs, we say we have a *success*. If, then, we are concerned with a die falling a 3, each time it falls a 3 we have a success. (Paradoxically, if we are concerned with the people who fail an examination, every failure is a success!!) The probability of a success is symbolised as p and so we can say that in the case of a die falling a 3, $p = \frac{1}{6}$.

If the event with which we are concerned does *not* occur we have, of course, a *failure*. The probability of a failure is symbolised as q, and a moment's thought will quickly show that $q = 1 - p$, i.e. that the probability of a failure is always 1 minus the probability of a success.

STRING-EVENT PROBABILITIES

The probability of any given elementary event is usually a fact of life and requires no calculation (e.g. the probabilities of tossing a head, throwing a 3, drawing an ace, having an accident, hitting a bull's-eye, over-sleeping, etc. cannot be calculated but must be found from observation or by logical deduction). On the other hand, the probability of a string-event can be calculated from the probabilities of the individual elementary events comprising the string. This section outlines the rules for these calculations.

4. The four-fold classification of string-event probabilities. There are four basic situations in which string-events are involved and which require different rules to find the string-event probabilities.

These are:

(*a*) *The "and/inclusive" situation.* In this situation we ask, "What is the probability of having this *and* that?" (where this and that are *inclusive* events).

(*b*) *The "or/exclusive" situation.* In this situation we ask, "What is the probability of having this *or* that?" (where this and that are *exclusive events*).

(*c*) *The "or/inclusive" situation.* In this situation we ask, "What is the probability of having this *or* that?" (where this and that are *inclusive* events).

(*d*) *The "and/exclusive" situation.* In this situation we ask, "What is the probability of having this *and* that?" (where this and that are *exclusive* events). This class is quickly disposed of since by being exclusive this and that *cannot* occur together and so the combination is impossible (i.e. probability = 0).

This four-fold classification, with memory symbols relating to the main features of the rules, is shown in Fig. 28 and those rules are detailed below.

	INCLUSIVE	EXCLUSIVE
'AND'	X	O
'OR'	List, X,+	+

FIG. 28 *Basic probability situations.*

5. The "and/inclusive" ("and") situation. In this situation we ask, "What is the probability of this *and* that?" (where this and that are inclusive events). Since the "and/exclusive" situation is impossible, we will henceforth refer to this situation simply as the "and" situation.

The rule here is very simple—merely *multiply* the elementary event probabilities. So the probability of tossing a head and throwing a 3 is $\frac{1}{2} \times \frac{1}{6} = \frac{1}{12}$, i.e. probability H−3 $= \frac{1}{12}$.

There is, incidentally, no limit to the number of elementary events that can enter the string-event. Thus, the probability that

A will toss a head *and* B will toss a head *and* C will toss a head *and* D will throw a 3 *and* E will throw a 3 = $\frac{1}{2} \times \frac{1}{2} \times \frac{1}{2} \times \frac{1}{6} \times \frac{1}{6} = \frac{1}{288}$. So we would say the probability of string-event H—H—H—3—3 = $\frac{1}{288}$.

6. The "or/exclusive" situation. In this situation we ask, "What is the probability of this *or* that?" (where this and that are *exclusive* events).

The rule here is equally simple—merely *add* the elementary event probabilities. So the probability of throwing a 3 *or* throwing a 4 with a single throw of a die (which are exclusive events since they cannot occur together) is $\frac{1}{6} + \frac{1}{6} = \frac{1}{3}$. Again, there is no limit to the number of elementary events that can enter into the string-event. Thus the probability of a single throw giving 1 *or* 2 *or* 3 *or* 4 *or* 5 *or* 6 = $\frac{1}{6} + \frac{1}{6} + \frac{1}{6} + \frac{1}{6} + \frac{1}{6} + \frac{1}{6} = 1$ (which makes sense since it is certain that one of the six faces will fall uppermost).

7. The "or/inclusive" situation. In this situation we ask, "What is the probability of this *or* that?" (where this and that are *inclusive* events).

NOTE: In probability theory by "or" we normally mean one or other *or both.*

This situation leads to much more complex rules. At first it may be thought that again one should do no more than add the probabilities. However, the error of this can be seen if one tries to find the probability of obtaining a 6 with a throw of seven dice (inclusive events since the 6 can turn up on more than one die), for adding gives us $\frac{1}{6} + \frac{1}{6} + \frac{1}{6} + \frac{1}{6} + \frac{1}{6} + \frac{1}{6} + \frac{1}{6} = 1\frac{1}{6}$—which is obviously nonsense since not only is a probability greater than 1 (certainty) impossible but it is also far from certain that there will, in fact, be a die showing a 6.

The rules for this situation, it turns out, call for a 3-stage procedure. These three stages are as follows:

(*a*) First, list *all* the possible string-events which give the result for which you are looking—i.e. give a "success."

(*b*) Next, find the probability of each string-event. Since a string-event is composed of elementary events in the "and" situation, this simply involves multiplying all the elementary event probabilities together.

(*c*) Finally, since all the string-events are exclusive and since what we really want to know is the probability of one of them occurring, we have an "or/exclusive" situation—which calls for the addition of all the string-event probabilities.

EXAMPLES

1. What is the probability of tossing a head or throwing a 4?

SOLUTION

(*a*) String-events giving a success: H–4, H–Not 4, Not H–4.

(*b*) Probabilities of each string-event (Note: Probability of Not 4 = Failure to throw 4 = $q = 1 - p = 1 - \frac{1}{6} = \frac{5}{6}$):

String-event	*Probability*
H–4 = H *and* 4	$\frac{1}{2} \times \frac{1}{6} = \frac{1}{12}$
H–Not 4 = H *and* Not 4	$\frac{1}{2} \times \frac{5}{6} = \frac{5}{12}$
Not H–4 = T *and* 4	$\frac{1}{2} \times \frac{1}{6} = \frac{1}{12}$

(*c*) Probability of H–4 *or* H–Not 4 *or* Not H–4 $= \frac{1}{12} + \frac{5}{12} + \frac{1}{12} = \frac{7}{12}$. So probability of tossing a head or throwing a 4 $= \frac{7}{12}$.

2. Three men have bets in three different races. A has a probability of winning of $\frac{1}{3}$, B a probability of $\frac{1}{4}$ and C of $\frac{2}{5}$. They agree that if any of them wins they will have a party. What is the probability of the party?

SOLUTION

(*a*) Coding a win as W and a loss as L (and remembering that the probability of a loss is 1 – probability of a win), the string-events that will give a success (i.e. result in a party) are:

AW–BW–CW, AW–BW–CL, AW–BL–CW, AL–BW–CW, AW–BL–CL, AL–BW–CL, AL–BL–CW.

NOTE

(*i*) In the "or/inclusive" situation it is as important to detail in the string-events the *losing* elementary events as it is the winning elementary events. So every string-event *must* record a possible outcome of every elementary event involved in the situation.

(*ii*) As will be appreciated, listing all the successful string-events in itself calls for careful thought.

(*b*) String-event probabilities:

	String-event *A*	*B*	*C*	*Probability*
1	W	W	W	$\frac{1}{3} \times \frac{1}{4} \times \frac{2}{5} = \frac{2}{60}$
2	W	W	L	$\frac{1}{3} \times \frac{1}{4} \times \frac{3}{5} = \frac{3}{60}$
3	W	L	W	$\frac{1}{3} \times \frac{3}{4} \times \frac{2}{5} = \frac{6}{60}$
4	L	W	W	$\frac{2}{3} \times \frac{1}{4} \times \frac{2}{5} = \frac{4}{60}$
5	W	L	L	$\frac{1}{3} \times \frac{3}{4} \times \frac{3}{5} = \frac{9}{60}$
6	L	W	L	$\frac{2}{3} \times \frac{1}{4} \times \frac{3}{5} = \frac{6}{60}$
7	L	L	W	$\frac{2}{3} \times \frac{3}{4} \times \frac{2}{5} = \frac{12}{60}$
			TOTAL	$= \frac{42}{60}$

(*c*) Probability of string-event 1 *or* 2 *or* 3 *or* 4 *or* 5 *or* 6 *or* 7 = Sum of string-event probabilities = Total as above = $\frac{42}{60}$ (= 0.7)

∴ Probability of party = 0.7.

8. The "at least" situation. There is another probability situation which, however, is more of a short-cut device than a new situation. This is the situation when we ask, "What is the probability that *at least x* successes will be achieved?", e.g. what is the probability that at least one 6 will fall if seven dice are all thrown at once?

Now adopting the "or/inclusive" rules above would require us to list all the string-events which will give us at least one 6, find by multiplication the probabilities of each and then add to find the answer probability. However, if the student chooses to list all these string-events he will find there are 127 of them!

Clearly, a short-cut would be preferable and such a short-cut does, in fact, become apparent once the link between the probability of an "at least" probability and certainty is appreciated. In the case here, for instance, it is certain that the seven dice will have at least one 6 or that *none of them will have a 6*, i.e.:

Probability of (at least one 6 *or* probability of no 6's at all) = Certainty.

i.e. Probability of at least one 6 + probability of no 6's = 1

∴ Probability of at least one 6 = 1 − probability of no 6's.

Now the only string-event that gives no 6's is the one where every die is not a 6, and the probability of that string-event is $\frac{5}{6} \times \frac{5}{6} \times \frac{5}{6} \times \frac{5}{6} \times \frac{5}{6} \times \frac{5}{6} \times \frac{5}{6} = \frac{78125}{279936} = 0.279$. So the probability of at least one 6 = 1 − 0.279 = 0.721.

This formula can be more generally stated as follows:

Probability of *at least x* successes =
1 − probability of less than *x* successes.

CONDITIONAL PROBABILITY

So far we have looked at situations where the probability of the success of any given elementary event does not change. This, of course, is not always the case. For instance, the probability of drawing the ace of spades out of a pack from which all the hearts have been removed is different from the probability of drawing that ace from a full pack.

In this section we see how our basic probability theory handles such probability changes.

9. Meaning of conditional probability. We have just pointed out that the probability of drawing the ace of spades from a pack of depends upon whether the pack is a full one or not. In other words, the probability *is conditional upon* the pack from which it is drawn. When the probability of an event depends upon a prior condition we say that the probability is a *conditional probability*.

10. Conditional probability calculations. Conditional probability calculations are sometimes arithmetically lengthy but they do not really involve any new probability concepts. Thus, if we have a full pack of cards the probability of drawing the ace of spades is $\frac{1}{52}$, while if we have a pack from which the hearts have been removed (leaving a total of only 39 cards) the probability is $\frac{1}{39}$. If we want to know the probability of drawing an ace of spades from a full pack and at the same time drawing another ace of spades from a second heartless pack we would simply adopt the "and" rule and say the probability was $\frac{1}{52} \times \frac{1}{39}$.

11. P(B | A). To avoid cumbersome expressions we code a conditional probability in the form P(B | A). Here "P" stands for "The probability of", "B" for the event in which we are interested, "|" for "given" and "A" for the condition imposed. So "P(drawing the ace of spades | heartless pack) = $\frac{1}{39}$" would read, "The probability of drawing the ace of spades given heartless pack = $\frac{1}{39}$".

12. Conditional probability string-events. Conditional probability becomes interesting where there are string-events. Imagine, for instance, we have an urn containing 10 marbles, 3 of which are red. What is the probability that the first two marbles drawn out are red?

NOTE: By tradition urns are regarded as indispensable to the study of probability.

Now what we are asking, of course, is, "What is the probability of the string event R–R?" So we need the "and" rule. Clearly, the probability of the first marble being red is $\frac{3}{10}$. However, note that after the first marble has been drawn there will only be 9 marbles left in the urn—and only 2 of those will be red. So the probability of the second marble being red is $\frac{2}{9}$. Our string-event R–R, then, has a probability of $\frac{3}{10} \times \frac{2}{9} = \frac{6}{90} = 0.067$.

As can be seen, this problem only differs from the ones we looked at in **11** in so far that the probability of drawing a red marble on the second draw, given the first marble drawn is red, is *not* the same as the probability of the first marble drawn being red. Nor, in fact, is it the same as the probability of the second marble being red, given that the first marble was not red. For in this case, after the first not-red marble has been drawn there will be 9 marbles in the urn, of which 3 will be red—so the probability of a red marble will be $\frac{3}{9}$.

Symbolically we would, of course, show the two different probabilities of this second marble being red as:

P(2nd marble red | 1st marble red) = $\frac{2}{9}$
P(2nd marble red | 1st marble not red) = $\frac{3}{9}$

This really is all there is to conditional probability theory, though more involved situations do result in more involved layouts.

EXAMPLE

In the case of the urn above, what is the probability that at least 2 of the first three marbles drawn will be red?

SOLUTION

This is an "or/inclusive" situation, so we apply the rules that lay down that the string-events which give a success must first be fully listed, the probability of each computed and these probabilities added. The layout below covers the whole of this procedure (it should be appreciated that finding the elementary event probabilities, is simply a matter of counting the relevant marbles in the urn at the time of any specific draw).

String-event			*Probability*
1st	2nd	3rd	
R	R	R	$\frac{3}{10} \times \frac{2}{9} \times \frac{1}{8} = \frac{6}{720}$
R	R	Not R	$\frac{3}{10} \times \frac{2}{9} \times \frac{7}{8} = \frac{42}{720}$
R	Not R	R	$\frac{3}{10} \times \frac{7}{9} \times \frac{2}{8} = \frac{42}{720}$
Not R	R	R	$\frac{7}{10} \times \frac{3}{9} \times \frac{2}{8} = \frac{42}{720}$
			TOTAL $= \frac{132}{720} = 0.183$

So the probability of drawing at least 2 red marbles in the first three marbles drawn is 0.183.

13. Definitions: prior event and subsequent event. Using conditional probability theory it is important that we distinguish between earlier events and later events, since the conditional probability of the later events depends upon just what earlier events occurred. Logically we define the earlier event as the *prior event* and the later event as the *subsequent event*. In our conditional probability symbol P(B | A), then, B stands for the subsequent event and A for the prior event.

14. P(B | any A). Very often we are interested in the probability of a particular subsequent event and are indifferent to whatever prior events may have preceded it. This, however, creates no problem—all it calls for is a listing of all the string-events which *terminate* with the required subsequent event and then the application of the rest of the normal "or/inclusive" rules.

EXAMPLE

My wife is to meet me by a given time. The probability of her being punctual if she travels by bus is 0.2, if she travels by train 0.5 and if she is given a lift in a neighbour's car 0.9. The probabilities of her travelling by bus, train and car are 0.3, 0.4 and 0.3 respectively. What is the probability that she will be on time?

SOLUTION

Here we are indifferent as to just how my wife travels—we are only interested in whether she will arrive on time or not. So all the relevant string-events are: Bus–arrive on time; train-arrive on time; car–arrive on time. And applying the "or/inclusive" procedure (since we are asking what is the probability she will travel by bus *and* arrive on time, *or* travel by train *and* arrive on time, *or* travel by car *and* arrive on time), we have:

String-event	*Probability*
Bus–arrive on time	$0.3 \times 0.2 = 0.06$
Train–arrive on time	$0.4 \times 0.5 = 0.20$
Car–arrive on time	$0.3 \times 0.9 = 0.27$
	TOTAL $= 0.53$

So the probability that she will arrive on time is just over a half.

15. Bayes' Theorem. There is an interesting extension of the technique developed in **14** above, since where there is more than one possible prior event and the subsequent event actually proves to be a success, we can ask ourselves what is the probability a given possible prior event did, in fact, occur.

Now note that this is *not* to ask simply what is the probability of a given prior event but what is the probability of a given prior event, *in view of the fact that the subsequent event occurred.* Thus, in the paragraph above, before my wife arrives the probability of her using a given mode of transport are the 0.3, 0.4 and 0.3 previously detailed. However *if she arrives on time* then we have an additional piece of information that changes the probabilities as to which mode she *actually* used.

Given, then, that she arrives on time, how can we compute the probability of each prior event having actually occurred? The answer to this lies in observing how frequently she would use each mode of transport on the successful occasions—i.e. on the occasions she arrives on time. (So that, for example, if she were always late when she walked, then if she arrived on time this very fact would have meant that the probability of her having walked must be "0" regardless of what the probability of her walking might have been before she set out.) Now as can be seen in the example layout in **14**, since there is a 0.06 probability of her travelling by bus *and* arriving on time then this event will occur 6 times in every 100 trips. Similarly, she will travel by train and arrive on time 20 times in every 100 and by car and arrive on time 27 times in every 100. And this means that out of the 53 times she actually arrives on time she will have travelled by bus 6 times, by train 20 times and by car 27 times. So the probability that, having arrived on time, she actually travelled by bus is $\frac{6}{53}$, by train is $\frac{20}{53}$ and by car $\frac{27}{53}$. (Note, incidentally, that while before setting out she has only a 0.3 probability of having a lift with the neighbour, if she arrives on time she is more likely to have had a lift than not.)

This method can be put in formula terms as follows:

Given that a particular subsequent event S has actually occurred the probability that a given possible prior event E actually occurred =

$$\frac{\text{Probability of prior event } E \times \text{conditional probability of subsequent event } S \text{ given } E}{\text{Probability of subsequent events occurring one way or another}}$$

or, P(Prior event E | subsequent event S)

$$= \frac{\text{Probability of } E \times P(S \mid E)}{\text{Probability of } S \text{ occurring one way or another}}$$

And this formula is known as *Bayes' Theorem.*

EXAMPLE

There are three urns—one black containing 3 red and 7 white marbles, one red containing 4 yellow and 6 green marbles and one white containing 8 yellow and 2 green marbles. A marble is first drawn from the black urn. A second marble is then drawn from the urn having the same colour as the first marble drawn. If the second marble is yellow, what is the probability that the first marble was red?

SOLUTION

Here the actual subsequent event is a yellow marble. There are only two string-events which can give this result—the R–Y and the W–Y events. The probabilities, then, of the second marble being yellow are:

String-event	*Probability*
R – Y (i.e. black and red urns)	$\frac{3}{10} \times \frac{4}{10} = \frac{12}{100}$
W – Y (i.e. black and white urns)	$\frac{7}{10} \times \frac{8}{10} = \frac{56}{100}$
∴ Probability of 2nd marble yellow	$= \frac{68}{100}$

∴ Probability (1st marble red | 2nd marble yellow)

$$= \frac{\text{Probability of R} \times P(Y \mid R)}{\text{Probability of 2nd marble Y}} = \frac{0.3 \times 0.4}{0.68} = 0.176$$

In other words, out of the 68 times (in 100) a yellow marble would be drawn second, on only 12 occasions would a red marble have been drawn previously from the black urn. So the probability that the first marble drawn was red is $\frac{12}{68} = 0.176$.

PROGRESS TEST 17

(Answers in Appendix VI

1. What is the probability of:

(*a*) (*i*) Tossing two heads,

(*ii*) Throwing a double six.

(*b*) Throwing a die and scoring a 5 or 6.

(*c*) Drawing an ace from a pack of cards.

(*d*) Tossing a head and a tail.

(*e*) Throwing two dice and scoring a total of 11.

(*f*) Throwing two dice and scoring 7.

(*g*) Backing at least one winner when betting on four horses having winning probabilities of $\frac{1}{3}$, $\frac{2}{5}$, $\frac{13}{40}$ and $\frac{4}{9}$.

2. The probability that a man now aged 55 years will be alive in 1993 is $\frac{5}{8}$, while the probability that his wife now aged 53 years will be alive in 1993 is $\frac{5}{6}$. Determine the probability that in 1993:

(*a*) both will be alive;

(*b*) at least one of them will be alive;

(*c*) only the wife will be alive. (*I.C.M.A.*)

3. In the past, two building contractors, A and B, have competed for 20 building contracts of which 10 were awarded to A and 6 were awarded to B. The remaining 4 contracts were not awarded to either A or B. Three contracts for buildings of the kind in which they both specialise have been offered for tender.

Assuming that the market has not changed, find the probability that:

(*a*) A will obtain all three contracts;

(*b*) B will obtain at least one contract;

(*c*) Two contracts will not be awarded to either A or B;

(*d*) A will be awarded the first contract, B the second, and A will be awarded the third contract. (*I.C.M.A.*)

4. The probability of my arriving at work less than 5 minutes late is 0.7 and of arriving 5 or more minutes late is 0.2. If I am less than 5 minutes late the probability of a reprimand is 0.4 while if I am 5 or more minutes late it is 0.9.

(*a*) What is the probability that I will avoid a reprimand tomorrow?

(*b*) If I got a reprimand today, what is the probability that I was less than 5 minutes late?

5. The probability that my son will use my car while I am at work is 0.3. The probability that my monthly garage bill will be reasonable if he does not use my car is 0.2, while if he does it will be 0. The bill has been unreasonable for the past three months. What is the probability that my son has, on occasions, used my car during this period?

CHAPTER XVIII

Probability Distributions

INTRODUCTION

In this chapter we outline very briefly the two most common probability distributions (i.e. after the normal curve, *see* Chapter XIX) — the binomial and the Poisson distributions.

1. What is a probability distribution? A *probability distribution* is very similar to a frequency distribution except that probabilities are used instead of frequencies. This is only a very minor difference since a probability distribution can be converted into a frequency distribution by simply multiplying all the probabilities by a total frequency figure. Thus, if we tossed two coins the probability of getting 0 heads would be the probability of getting two tails, which is $\frac{1}{2} \times \frac{1}{2} = \frac{1}{4}$; the probability of getting a head would be the probability of getting a head and a tail or a tail and a head, which is $(\frac{1}{2} \times \frac{1}{2}) + (\frac{1}{2} \times \frac{1}{2}) = \frac{1}{2}$; and the probability of getting two heads is $\frac{1}{2} \times \frac{1}{2} = \frac{1}{4}$. If we lay out these probabilities in the following form:

No. of Heads	*Probability*
0	0.25
1	0.50
2	0.25
	1.00

we have the probability distribution relating to tossing two coins. Obviously, if we tossed two coins for a total of 100 tosses we would expect the following results:

No. of Heads	*Probability*	*Frequency*	
0	0.25	0.25 × 100 =	25
1	0.50	0.50 × 100 =	50
2	0.25	0.25 × 100 =	25
	1.00		100

which is, of course, a frequency distribution.

2. Probability distribution symbols. The probability distributions that are generally the most useful are those that can be expressed as one-line formulae. The most common symbols for such formulae are the following (which should be memorised):

p = probability of a success—which in this case means the probability that an item selected at random has a required characteristic:

(e.g. The probability that a red marble is selected from an urn containing red and white marbles.)

%

$$q = \text{probability of failure} = 1 - p$$

n = number of items selected from the total population.
P(x) = probability that exactly x items out of the n selected will have the required characteristic:

(e.g. P(3) would symbolise the probability that exactly 3 red marbles would be in our selection of n items.)

$$\binom{n}{x}^* = \frac{n!}{(n-x)!\,x!}$$

* To understand this symbol one really needs to understand the mathematics of combinations. However, all calculations will prove correct if the formula given is followed. Note that "!" is called the *factorial* sign and means that one must list in descending order all the numbers from and including the number that precedes the ! down to 1—and then multiply the whole lot together (e.g. $6! = 6 \times 5 \times 4 \times 3 \times 2 \times 1 = 720$). Note, too, that $0! = 1$. Odd, but true.

THE BINOMIAL DISTRIBUTION

Imagine we have an urn containing 100 000 marbles, 40,000 of which are red. A random sample of 4 is taken. What are the probabilities that in this sample there are 0 red marbles, 1 red marble, 2 red marbles, 3 red marbles and 4 red marbles respectively?

Using our conditional probability theory we would begin to solve this problem by saying, that since 40 000 marbles were red and 60 000 were not red then:

Probability of 0 reds, i.e. P(0)

$$= \frac{60\,000}{100\,000} \times \frac{59\,999}{99\,999} \times \frac{59\,998}{99\,998} \times \frac{59\,997}{99\,997}$$

Probability of 1 red, i.e. P(1)
= probability of 1st marble selected being only red marble + probability of 2nd marble being only red marble + etc.

And the probability of 1st marble only being red

$$= \frac{40\,000}{100\,000} \times \frac{60\,000}{99\,999} \times \text{etc.}$$

By this point the student has doubtless decided that life is too hort to go through the whole exercise and that if other people ave nothing better to do, he certainly has.

. **The binomial distribution—a short-cut.** Fortunately, however, he exercise need not be entirely abandoned for there is a relaively simple formula that gives a very good approximation to the rue answer. This formula is:

$$P(x) = \binom{n}{x} p^x q^{n-x}$$

In other words, the probability of exactly x items in a sample f n having the required characteristic equals $\frac{n!}{(n-x)!\,x!}$ time he probability that 1 item selected at random has the required haracteristic to the power of x, times $(1 - p)$ to the power of $n - x$). All we need to do now is to slot the actual values we ave for n, p and q into the formula and then compute $P(x)$ for ach required value of x.

This formula gives what is called the *binomial distribution.*

. **An illustrative binomial distribution.** To see how this formula vorks let us take our marble problem and find the probability istribution that we want.

First it should be noted that in this case

$$n = 4, p = \frac{40000}{100000} = 0.4, \text{ and so } q = 1 - 0.4 = 0.6$$

he binomial distribution formula becomes:

$$P(x) = \binom{4}{x} 0.4^x\, 0.6^{4-x}$$

or, since

$$\binom{4}{x} = \frac{4 \times 3 \times 2 \times 1}{(4 - x)!\, x!} = \frac{24}{(4 - x)!\, x!}$$

then

$$P(x) = \frac{24}{(4 - x)!\, x!} \times 0.4^x\, 0.6^{4-x}.$$

Now what we want are the probabilities of 0 red, 1 red, 2 red 3 red and 4 red marbles in our sample of 4. So x must be 0, 1, 2 3 and 4 in turn. We obtain, then, the following layout:

x (No. red marbles)	P(x) (i.e. probability of x red marbles in sample)
0	$\frac{24}{(4 \times 3 \times 2 \times 1) \times 0!^*} \times 0.4^{0}\dagger \times 0.6^{4-0}$ $= 1 \times 1 \times 0.6^4 = 0.1296$
1	$\frac{24}{(3 \times 2 \times 1) \times 1!} \times 0.4^1 \times 0.6^{4-1}$ $= 4 \times 0.4 \times 0.6^3 = 0.3456$
2	$\frac{24}{(2 \times 1) \times (2 \times 1)} \times 0.4^2 \times 0.6^{4-2}$ $= 6 \times 0.4^2 \times 0.6^2 = 0.3456$
3	$\frac{24}{1! \times (3 \times 2 \times 1)} \times 0.4^3 \times 0.6^{4-3}$ $= 4 \times 0.4^3 \times 0.6^1 = 0.1536$
4	$\frac{24}{0! \times (4 \times 3 \times 2 \times 1)} \times 0.4^4 \times 0.6^{4-4}$ $= 1 \times 0.4^4 \times 0.6^{0}\dagger = 0.0256$
	1.0000

So we now have a table showing how the probabilities are "distributed" in our specific situation.

5. When a binomial distribution can be used. We said in 3 that the binomial distribution was an *approximation*. The question, therefore, arises as to just when we are allowed to use this useful short-cut. The answer is that the binomial distribution can be used *whenever the value of* p *remains virtually unchanged during the time*

* Remember, 0! = 1.

† Note that *any* number to the power of 0 always equals 1.

the sample is being taken. In the case above, the probability of selecting a red marble stayed virtually unchanged at 0.4, since taking 4 marbles from a population of 100 000 hardly affects the probability at all.

One interesting point arises in respect of the use of the binomial distribution and that is that if the *population is infinite* p *never changes* and so the binomial distribution gives the *correct* probabilities—not merely an approximation. Infinite populations exist where there is coin tossing or dice throwing since, for example, a toss of two coins represents a single sample taken from an infinite population of potential two-coin tosses.

6. Mean and standard deviation of the binomial distribution. We state, without proof (though the student can make checks on actual distributions if he wishes), that the mean and standard deviation of a binomial distribution are:

$$\text{Mean} = np$$
$$\text{Standard deviation} = \sqrt{npq}$$

EXAMPLE

Taking the distribution in **4**:
Mean = 4 × 0.4 = 1.6 red marbles (*i.e.* if thousands and thousands of samples of 4 marbles were taken, then the average number of red marbles per sample would be 1.6).
Standard deviation = $\sqrt{4 \times 0{\cdot}4 \times 0{\cdot}6}$ = 0·980 red marbles.

THE POISSON DISTRIBUTION

Next let us imagine that in our 100 000-marble urn there are now only 120 red marbles but that our sample size is to be 1000. What now are the probabilities of having 0, 1, 2, 3 and 4 red marbles in our random sample?

In this instance p is obviously only $\dfrac{120}{100\,000} = 0{\cdot}0012$ and so employing the binomial distribution will involve using the formula

$$P(x) = \frac{1000!}{(1000 - x)!\,x!} \times 0{\cdot}0012^{x} \times 0{\cdot}9988^{1000-x},$$

where x takes every value from 0 to 1000. Even in the simplest case, where x is only 0, we will be called upon to find $0{\cdot}9988^{1000}$.

Clearly, another kind of short-cut is very much wanted.

7. The Poisson distribution. Fortunately, a short-cut exists since in these circumstances the *Poisson distribution* gives us a very good approximation to the binomial distribution. The formula for the distribution is:

$$P(x) = e^{-a} \times \frac{a^x}{x!}$$

where a = average number of items per sample.

At first sight this looks rather complicated since we still have to find a and then compute e^{-a}, whatever e may happen to be. However, these problems are trivial. It turns out that a can be found from the formula $a = np$, and that e is a constant (rather like π) which means that a table for e^{-a} can be prepared so that finding e^{-a} for any value of a simply involves looking up the table—and such a table is given in Appendix V (*b*).

EXAMPLE

Taking our particular illustration, $p = 0.0012$ and $n = 1000$, so $a = 0.0012 \times 1000 = 1.2$. And Appendix V(*b*) shows us that $e^{-1.2}$ (i.e. e^{-a}) = 0.3012.

8. An illustrative Poisson distribution. Let us now apply the Poisson distribution to the specific case where a 1000-marble sample is selected from a large population in which the probability of selecting a red marble is 0.0012. As we saw in 7 above, in this case $a = 1.2$ and $e^{-a} = 0.3012$. This means that our Poisson distribution formula becomes:

$$P(x) = 0.3012 \times \frac{1.2^x}{x!}$$

Since we want the probability of having 0, 1, 2, 3, 4, etc. red marbles in our sample of 1000 the layout we need is as follows:

x (No. red marbles)	P(x) (i.e. probability of x red marbles in sample)
0	$0.3012 \times \frac{1.2^0}{0!} = 0.3012$
1	$0.3012 \times \frac{1.2^1}{1!} = 0.3614$
2	$0.3012 \times \frac{1.2^2}{2!} = 0.2169$
3	$0.3012 \times \frac{1.2^3}{3!} = 0.0867$

Dear Reader,

We hope you found this M & E Handbook a valuable aid for your studies.

If you would be kind enough to complete and return this questionnaire (postage is prepaid in the UK) it would help us in our continuing efforts to provide books that are fairly priced, readily available, on appropriate subjects and at the right level.

1. Have you bought M & E Handbooks before?
- ☐ a) Yes
- ☐ b) No

2. Why did you buy this book? Because:
- ☐ a) it was recommended reading
- ☐ b) you saw it advertised
- ☐ c) you saw it displayed in the shop
- ☐ d) you heard about it from a friend
- ☐ e) other reason

3. In your area, do you find M & E Handbooks:
- ☐ a) easy to obtain
- ☐ b) limited selection available locally
- ☐ c) difficult to obtain

4. In which town did you purchase this book:

----

5. Do you find M & E Handbooks in your local stockist are:
- ☐ a) well displayed
- ☐ b) poorly displayed
- ☐ c) only available to order

6. Do you think this Handbook is fairly priced?
- ☐ a) Yes
- ☐ b) No

7. Where are you studying?
- ☐ a) College of Further Education/Technical College
- ☐ b) School
- ☐ c) University/Polytechnic
- ☐ d) College of Higher Education/Technology
- ☐ e) Other

8. Is your course of study
- ☐ a) Full time
- ☐ b) evening classes
- ☐ c) a short course of very limited duration
- ☐ d) part time/day release
- ☐ e) postal tuition
- ☐ f) other

9. In which subject areas would you like to see more M & E Handbooks published?

----

10. What exams are you studying for?

Thank you for completing this questionnaire, now simply remove the card from the book and drop the card into a letter-box. (No postage stamp necessary if posted in UK.)

Postage
will be
paid by
licensee

Do not affix Postage Stamps if posted in
Gt Britain, Channel Islands, N Ireland
or the Isle of Man

BUSINESS REPLY SERVICE
Licence No WC3594

Macdonald & Evans
Estover,
Plymouth,
PL5 2BR.

4	$0.3012 \times \frac{1.2^4}{4!} = 0.0260$
5	$0.3012 \times \frac{1.2^5}{5!} = 0.0062$
6 1000	Since we know that the whole probability distribution must add to 1 (since it is certain there will be 0 or 1 or 2 or 3 or 4 . . . or 1000 red marbles in the sample) all these P(x)s must *collectively* be the balance required to make the total equal 1. $= 0.0016$
	1.0000

As can be seen, the probabilities drop rapidly once x passes 2, and that in fact the probability of a sample of 1000 containing more than 5 reds (let alone a specific number of reds) is a mere 0.0016.

9. When a Poisson distribution can be used. Again, the Poisson distribution can only be used in particular circumstances (though those circumstances are very common). It can, in fact, be used *whenever the sample size is very large and the probability of a success is very small*, i.e. when n is very large and when p is very small. This wording of the restriction, of course, raises the problem of how large is "very large" and how small is "very small." As was said in 7, the Poisson distribution is an approximation to the binomial distribution and so the larger n and the smaller p the better the approximation will be. However, for an all-round rule the Poisson distribution can usually be used when p is less than 0.1 (when a sample size of less than 20 would in practice be hardly worth taking).

10. Mean and standard deviation of the Poisson distribution. Again we will state without proof the mean and standard deviation of the Poisson distribution. These are:

$$\text{Mean} = a$$
$$\text{Standard deviation} = \sqrt{a}$$

EXAMPLE

Taking the distribution in **8**:
Mean = 1.2 red marbles per sample
Standard deviation = $\sqrt{1.2}$ = 1.095 red marbles.

11. The Poisson distribution and an unknown sample size. The Poisson distribution has one very interesting (and useful) feature —it can be used even though the sample size is not known. This is because the Poisson formula only requires the *average* number of successes to be known and we can know the average without knowing either n or p. For example, to take the classical case, assume that we are studying thunderstorms and have been counting the number of lightning flashes. We find that over a total period of 100 minutes there were 120 flashes, i.e. an *average* of 1.2 flashes a minute. Now if we regard a particular minute as a very large sample of many very small moments of time and if there is a very small but constant probability of a single lightning flash occurring during each moment of time, then we can regard the pattern of the flashes as forming a Poisson distribution in which $a = 1.2$. So we can prepare a Poisson probability distribution that will tell us the probability of having in a given minute 0, 1, 2, 3, 4 etc. lightning flashes.

PROGRESS TEST 18

(*Answers in Appendix VI*)

1. It has been established that in a very large city 10% of the houses lack a bathroom.

(*a*) Find the probabilities that a random sample of 10 houses will contain 0, 1, 2, 3, 4, 5, 6 or more houses without bathrooms, using:

(*i*) The binomial distribution.
(*ii*) The Poisson distribution.

(*b*) Graph the two distributions on the same graph.

(*c*) From the binomial distribution find the probability that the sample will contain less than 2 bathroomless houses.

Work to 4 decimal places.

NOTE: This is very much a borderline situation for using a Poisson distribution. Nevertheless, comparison of this distribution with the binomial distribution (to which it approximates) will give the student some insight into the extent of error that arises when the Poisson distribution is used in lieu of the binomial distribution.

PART SIX

SAMPLING THEORY

CHAPTER XIX

The Normal Curve of Distribution

THE ROLE OF SAMPLES

1. From sample to population. So far, our statistical studies have concentrated on maximising our comprehension of the groups of figures we have collected. Now we go a step further: to maximise our comprehension, not just of the figures we collect, but of *the populations we collect them from.* Indeed, this is more a leap than a step, for consider what it implies. What we have been doing so far has been to take, say, a collection of horses in a field and find ways of describing the horses we see. Now we shall be looking at the horses in a field and, on the basis of that sample, describing *all* horses: including the thousands we cannot see.

2. Basic points. On the face of it, an attempt to formulate an all-embracing description from a tiny fraction of the total population can be no more than rash guesswork. But this is not so. There are a number of statistical laws which govern the kind of distribution usual in such work and which will ensure that our descriptions have a scientific basis. The most important of these laws—those relating to the normal curve of distribution—are examined in this chapter.

Two things, however, do need emphasising:

(*a*) Our samples *must* be random samples, otherwise we run the risk of serious error.

(*b*) We can never be absolutely certain our descriptions are correct. This means that every time we offer a description *we must indicate the probability of our description being in error.*

THE NORMAL CURVE OF DISTRIBUTION

3. Mathematical curves. Students may recall from their school algebra that curves drawn on a graph could be described by purely mathematical expressions such as $y = 10x^2 + 6x + 3$ and, because these curves were mathematical, facts about the graph could be *calculated.*

One important group of facts that can be calculated in this way concerns the area contained by the curve and other given parts of the graph.

The task of handling statistical data can often be simplified if a mathematical curve can be found approximating to that which would be produced by plotting the actual data on a graph. By substituting the mathematical curve for the real one, we can make calculations to reveal facts about the distribution of the raw data, facts that would otherwise have been difficult to determine.

4. The normal curve of distribution. Students who have drawn a number of histograms may have been struck by the frequent recurrence of a pattern in which there is a high column in the centre of the histogram, with decreasing columns spread symmetrically on either side. If the class intervals were small enough, the frequency curve probably looked rather like a cross-section of a bell. Now it so happens that this pattern does in fact occur frequently in statistical work. There also happens to be a mathematical curve very similar to it, called the *normal curve of distribution.* It has the following features (*see* Fig. 29):

(*a*) It is symmetrical.
(*b*) It is bell-shaped.
(*c*) Its mean lies at the peak of the curve.
(*d*) The two tails never actually reach the horizontal axis, although they continuously approach it.
(*e*) The formula for the curve is:

$$y = \frac{1}{\sigma\sqrt{2\pi}} e^{-\frac{1}{2}\left(\frac{x-\bar{x}}{\sigma}\right)^2}$$

but students may ignore it completely at this stage, as any mathematical data relating to the curve can easily be found in mathematical tables. What *is* important is that the mathematical properties of the curve apply equally to the actual curve of the "raw" data. In particular, areas of the graph that lie below the normal curve will correspond to areas below the actual curve.

NOTE the inclusion of σ in the formula. It is because σ forms part of the normal curve of distribution formula that the standard deviation is so important in statistics.

5. Areas below the normal curve of distribution. There is a distinct relationship between σ and the areas under a normal curve. This relationship is such that if we:

(*a*) erect a line at the centre of the curve (i.e. at the mean),
(*b*) measure off 1σ lengths either side of this central line,
(*c*) erect perpendicular lines at these points,

then the area enclosed between the curve, the 1σ vertical lines and the axis is 68.26 per cent of the total area enclosed by the whole curve (*see* Fig. 30). Of course, since the curve is symmetrical, the area between the central line and each 1σ line is half this value, i.e. 34.13 per cent.

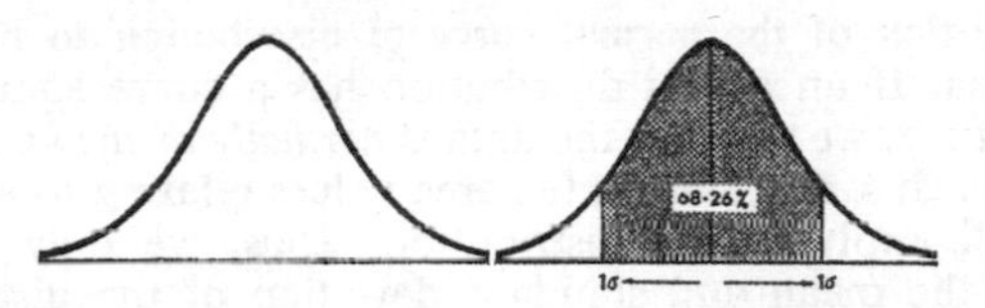

FIG. 29 *Normal curve of distribution.*

FIG. 30 *Area enclosed between the 1σ points.*

Mathematicians can, in fact, compute the area that lies between *any* two lines, the curve and the axis—always measuring along the axis in units of σ. Obviously a table of such areas is of considerable value, and one is given in Appendix IV. In this table the area between the mean line, the curve, the horizontal axis and a vertical line measured from the mean in units of σ, is given as a decimal of the total area enclosed by the curve. For instance, if the line lies $1\frac{1}{2}\sigma$ from the mean, the table shows that the area enclosed is 0.4332 (i.e. 43.32 per cent) of the total area.

6. Normal curve values to be learnt. The student is advised to learn three normal curve values: those relating to the approximate areas between each pair of 1σ, 2σ and 3σ lines as shown in Fig. 31. Note that almost the entire area lies within 3σ of the mean.

7. Areas and frequencies. At this point, students should cast their minds back to Chapter X, where we discussed histograms and frequency curves. There, it was emphasised that, on graphs of distributions, areas were proportional to frequencies. This means that an area of 95 per cent of the total area covers a range of the distribution that includes 95 per cent of the total frequencies. This means that *areas and frequencies are interchangeable*, i.e. x per cent of the area covers x per cent of the frequencies and vice versa.

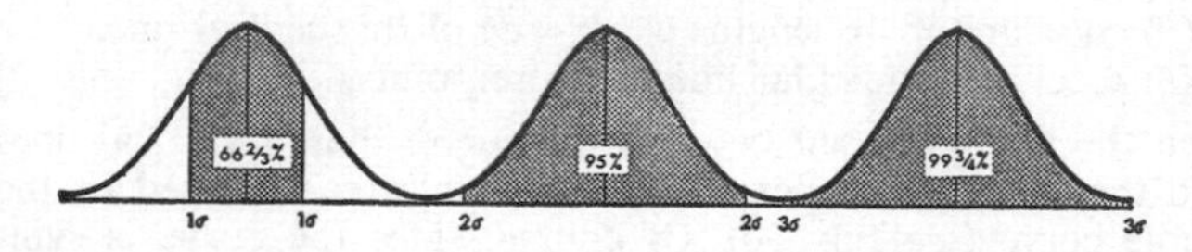

FIG. 31 *Approximate areas beneath the normal curve.*

8. Application of the normal curve of distribution to an actual distribution. If an actual distribution has a curve similar to a normal curve, we say that the data is *normally distributed.* When we have such a distribution the area values relating to a normal curve will apply to the distribution. Thus, we only have to compute the mean and standard deviation of the distribution in order to equip ourselves for determining how much of the distribution lies between any given points.

Let us now see what this means in terms of actual figures. Previously we found that Table VIII had a mean of $454\frac{1}{2}$ kilometres and a standard deviation of 27 kilometres (*see* Appendix III). Applying the area figures of the normal curve of distribution, this means that $66\frac{2}{3}$ per cent of the area (i.e. distances) will lie between the mean $\pm$ 1σ—which is $454\frac{1}{2} + 27$ km and $454\frac{1}{2} - 27$ km, i.e. $427\frac{1}{2}$ and $481\frac{1}{2}$ km. (These estimates can be checked from the ungrouped frequency distribution in Table VII, which reveals that 78 distances, i.e. 65 per cent of the 120 items, lie between these limits. This is not the $66\frac{2}{3}$ per cent expected, but the difference arises because the Table VIII distribution does not quite follow a normal curve—it is slightly skewed.) Similarly, 95 per cent will theoretically lie between the 2σ lines, i.e. between $454\frac{1}{2} + 2 \times 27$ and $454\frac{1}{2} - 2 \times 27 = 400\frac{1}{2}$ and $508\frac{1}{2}$ kilometres (although 120 distances really constitute too small a distribution for testing the theoretical figures).

9. Recognising a normal curve. Students are warned that normal curves of distribution are not always immediately recognisable. Basically, they are always bell-shaped, but the "bell" may be very flat and broad or tall and thin, as shown in Fig. 32. Study of V, 1 will indicate the reason for these extremes: they are due simply to the stretching or compressing of the scales.

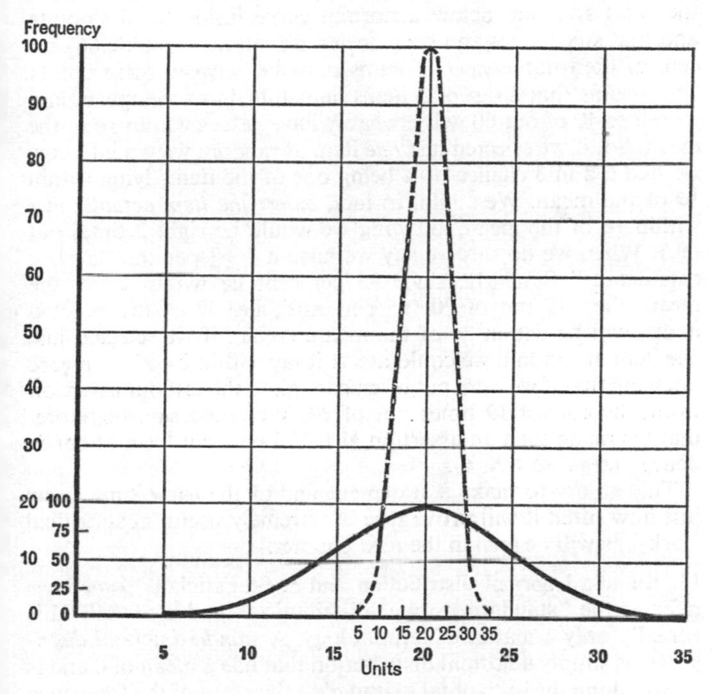

FIG. 32 *Normal curve of distribution on graphs of different scales.*

Here a normal curve having a mean of 20 units and a standard deviation of 5 units is plotted twice using different scales. The bold curve is the curve we obtain plotting on a graph with the bold type scales. The dotted curve arises if the light type scales are used. As can be seen, the appearances differ—yet both curves are the same normal curve.

Compare with Fig. 29, where the normal curve can be seen looking the usual shape.

10. Probability and the normal curve. Suppose that 60 items were selected *at random* from a large normal distribution: how many of the 60 do you think would have values that lay within 1σ of the mean? We know that approximately $66\frac{2}{3}$ per cent of the total area lies below a normal curve inside the 1σ points, and that since areas and frequencies are interchangeable $66\frac{2}{3}$ per cent of the total *number* of items must lie between these points. This means that 2 out of 3 items lie within 1σ of the mean, and therefore 40 of our 60 will probably have values within 1σ of the mean. So, if we selected just *one* item at random we would know we had a 2 in 3 chance of it being one of the items lying within 1σ of the mean. We could, in fact, *assert the item actually was* within 1σ of the mean, knowing we would be right 2 times out of 3. When we do this we say we have a "$66\frac{2}{3}$ per cent level of confidence." Similarly, since 95 per cent lie within 2σ of the mean, then 19 out of 20 (95 per cent), i.e. 57 of the selected items, will be within 2σ of the mean. Again, if we selected just *one* item at random we could assert it lay within 2σ of the mean, knowing that if we were put often enough to the test our assertion would be correct 19 times out of 20. We could say, therefore, that we made such an assertion at a "95 per cent level of confidence" (*see also* XX, **6**).

This ability to make a statement and at the same time know just how often it will prove true is extremely useful in statistical work—as will be seen in the next chapter.

11. Standard normal distribution and *z*. Statisticians sometimes refer to the "standard normal distribution" and also "*z*." This is really only a matter of terminology. A *standard normal distribution* is simply a normal distribution that has a mean of 0 and is scaled along the horizontal axis in σ's—the areas of the frequency curve of such a distribution being given in Appendix IV.

Now as we saw in **8**, it only requires commonsense to apply such a standard normal distribution to an actual distribution. What we in effect do, is to superimpose the actual distribution on the standard normal distribution scale so that the actual distribution mean coincides with the zero and the actual distribution values coincide with the appropriate σ points—i.e. our mean of $454\frac{1}{2}$ kilometres is at 0 and our $454\frac{1}{2} - 27$ and $454\frac{1}{2} + 27$ values lie at the -1σ and $+1\sigma$ points.

This is, of course, no more than a rather roundabout, though more mathematically rigid (and hence, in advanced statistics, a more useful), way of enabling us to state an actual value in terms of its deviation from the mean in units of standard deviation. And this new stated value is coded "z." As a formula, then, we have:

$$z = \frac{x - \bar{x}}{\sigma}$$

(NOTE: This is, in fact, the formula in Appendix IV.)

For example, take the actual value 400 km. From the formula we have

$$z = \frac{400 - 454\frac{1}{2}}{27} = -2{\cdot}02, \text{ i.e. the 400-km } z \text{ value is } -2{\cdot}02.$$

In other words, the 400-km value lies fractionally over two standard deviations below the mean of our distance distribution.

PROGRESS TEST 19

(*Answers in Appendix VI*)

1. What approximate percentage of the total area is enclosed by the normal curve of distribution, the central line, the axis and (*a*) the 2σ line, (*b*) the 3σ line and (*c*) the 2.6σ line?

2. What approximate area lies below the normal curve between:

(*a*) the lower 1σ and lower 2σ lines,
(*b*) the lower 1σ and upper 2σ lines,
(*c*) the lower 2σ and upper 3σ lines,
(*d*) the lower 2.6σ and upper 2.6σ lines?

3. What approximate area lies below the normal curve of distribution and *outside* (*a*) 1σ lines, (*b*) 2σ lines and (*c*) 3σ lines?

4. Assuming that the weight of 10 000 items are normally distributed and that the distribution has a mean of 115 kg and a standard deviation of 3 kg:

(*a*) How many items have weights between: (*i*) 115 and 118 kg, (*ii*) 112 and 115 kg, (*iii*) 109 and 121 kg, (*iv*) 106 and 124 kg?

(*b*) If you had to pick one item at random from the whole 10 000 items, how confident would you be in predicting that its value would lie between 109 and 121 kg?

CHAPTER XX

Estimation

So far all we have found is that many distributions are what is termed normal and that, knowing the mean and standard deviation, it is possible to say how many items lie between various limits. Our next step is to see how we can apply this knowledge to samples and populations, though in this and the next chapter we shall limit ourselves to what are called "large" samples—i.e. above 30 in size.

NOTE: The understanding of this chapter will be greatly improved if each time the word "mean" arises, the student thinks carefully as to just what is being said.

THE STANDARD ERROR OF THE MEAN

1. The distribution of a sample. Let us assume there is a population about which we know nothing except that it exists. Unknown to us, it is normally distributed, with a mean of 20 units and a standard deviation of 5 units. In order to learn something about this population we take a random sample and graph the distribution. How will it look?

We know that in any normal distribution 1 in every 3 items will lie between the lower 1σ and the mean, i.e. 15 and 20 in this case. Clearly, then, in any random sample approximately 1 out of every 3 items will also lie between 15 and 20, i.e. $33\frac{1}{3}$ per cent of the distribution of the sample. Similarly, approximately $33\frac{1}{3}$ per cent of the sample items will lie between 20 and 25, 14 per cent between 10 and 15, 14 per cent between 25 and 30, etc. (*see* Fig. 33). Now let us argue backwards and say that if a normal distribution has a particular pattern as regards areas, then any distribution having such a pattern must be normally distributed (remember, a normal distribution is simply one in which the frequency curve approximates to a particular mathematical curve). This means that the *distribution of our sample is also normally distributed* (occasionally, one-sided samples will be taken, but this will be rare). Moreover, since this distribution is such that 33⅓ per cent of the sample items lie between 15 and 20, etc. it is clear that our sample standard deviation will be 5 and our sample mean will be 20.

However, there is just one qualification to this. We ended the argument by assuming that the proportions of sample items between various limits matched those of the population exactly. In practice our sample will *not* be an exact miniature replica of the population—no sample ever is. Chance plays a big part in the selection of items in a sample; inevitably the sample mean and standard deviation will differ a little from the population mean and deviation (though the bigger the sample the less chance

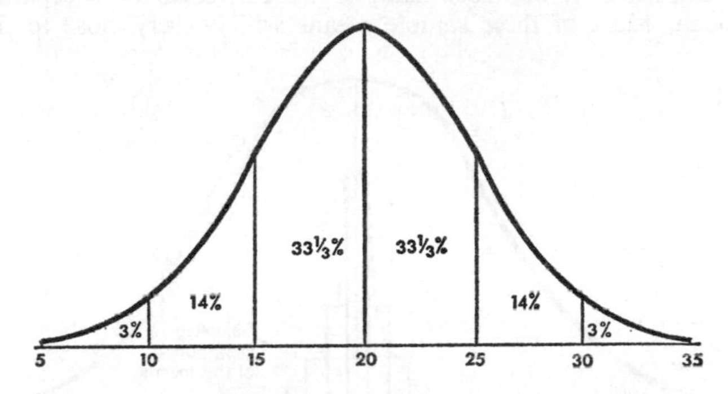

FIG. 33 *Distribution of a sample taken at random from a population with $\bar{x}$ 20 units and σ 5 units. Approximate percentages only.*

there is of selecting a sample which is unrepresentative of the population).

Our final conclusion, therefore, is that the mean and standard deviation of our sample will approximate to the mean and standard deviation of the population, i.e.

(*a*) the best estimate of the population mean is the sample mean, and

(*b*) the best estimate of the population standard deviation is the sample standard deviation.

2. Sampling distribution of the means. If we are ever required to estimate a population mean, we can take a sample and use the sample mean as the estimate. But although we know it will be close to the population mean, we do not know how great our error may be. What we need is some measure that will tell us the extent to which sample means will deviate from the population means. One approach to the problem is to examine just how far from the population mean the sample means of a large number of samples actually taken do, in fact, lie.

Let us assume that 2000 samples are taken, the size of each being 100 items.

NOTE: Students should distinguish very carefully between (*a*) the number of samples (2000) and (*b*) the number of items in a sample, i.e. *sample size* (100).

For each of our 2000 samples we can calculate a separate mean. Most of these sample means will be very close to the

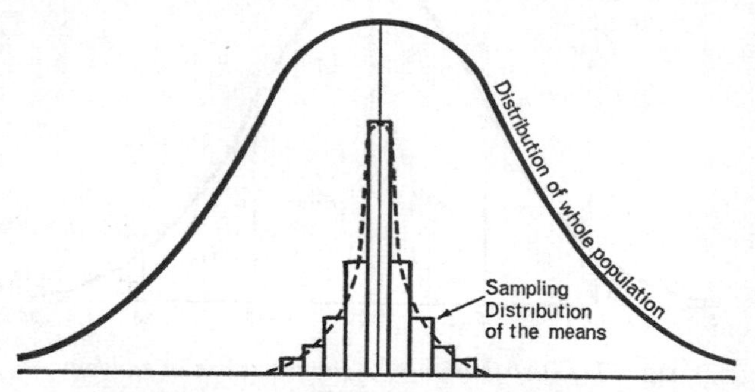

FIG. 34 *Sampling distribution of the mean.*

true mean of the population, though occasionally a sample will by chance contain an undue number of items with high (or low) values, so that its mean will be considerably above (or below) the population mean.

Next, we draw a histogram of the sample means (Fig. 34). On examination, the curve of this histogram looks suspiciously like a normal curve—a very "thin" one to be sure, but a normal curve for all that. Students will thus not be surprised to learn that this distribution of means *is* normally distributed. Moreover, it can be seen that the mean of such a distribution (i.e. the arithmetic mean of all the sample means) will be equal to the true population mean.

To summarise: if we graph the distribution of the *means* of a number of samples we find that it results in a normal curve with a mean equal to the population mean. Such a distribution is called a *sampling distribution of the mean.*

NOTE: Although for explanatory purposes we have assumed that the population distribution was normal this is not a necessary condition. The distribution can be badly skewed but our theory still holds.

3. The standard error of the mean. The next question is, what will be the standard deviation of such a distribution? Theoretically it *could* be found in the usual way, i.e. by taking the sample means as the variable and using the ordinary formulae for finding the standard deviation. But if this were done in practice the work involved would be so extensive that the result would hardly justify it. Fortunately, no such work is necessary, for it so happens that there is a connection between the standard deviation of the means, the standard deviation of the population, and the sample size. The connection is such that:

$$\sigma_{\bar{x}} = \frac{\textit{Standard deviation of the population}}{\sqrt{\textit{Sample size}}}$$

where $\sigma_{\bar{x}}$ is the standard deviation of the means, or the *standard error of the mean*, as it is called.

At first glance this does not seem to improve matters, as we still need to know the standard deviation of the population. However, we saw earlier that the standard deviation of the population was approximately equal to the standard deviation of a sample taken from that population. The formula can therefore be re-written as:

$$\sigma_{\bar{x}} = \frac{\textit{Standard deviation of the sample}}{\sqrt{\textit{Sample size}}}$$

This means that we need to take only *one sample* in order to find the standard deviation of a distribution of the means of a whole host of samples.

ESTIMATING POPULATION MEANS

4. The importance of the standard error. When we have found the standard error, what use is it? Simply this. Since 95 per cent of the items in a normal distribution lie within 2σ of the mean of the distribution, and since the distribution of the means is normal with a mean equal to the population mean, then 95 per cent of the means of *all* samples must lie within two standard errors of the true mean of the population.

This means that if we take a single sample, then 19 times out of 20 the sample mean will lie within two standard errors of the true mean of the population. In other words, 19 *times out of* 20 *the true mean of the population cannot lie more than two standard errors from the mean of the sample.*

5. Estimating the true mean. Let us look again at what we are saying from the beginning. We take a sample from a large population. We find the mean of the sample and know it must be close to the true mean of the population—so close in fact that we can say it is approximately the true mean.

Unfortunately, we do not know how great our error will be in taking the sample mean as the true mean. However, if we compute the standard error from the standard deviation of our sample, we shall be able to say that 19 times in 20 the true mean will be within two standard errors of the sample mean, i.e. the population mean lies within the "sample mean $\pm 2\sigma$" range.

EXAMPLE

Assume that our distances in Table VIII were but a random sample taken from a much larger population of distances, and that we wish to estimate the true mean distance of that population.

The mean distance of our sample, Table VIII, was 454½ kilometres and its standard deviation was 27 kilometres (*see* Appendix III).

$\therefore$ the best estimate of the mean and standard deviation of the population is also 454½ and 27 kilometres.

$$\therefore \text{ standard error } (\sigma_{\bar{x}}) = \frac{\sigma_x}{\sqrt{n}} = \frac{27}{\sqrt{120}} = 2.46 \text{ kilometres}$$

Now, 19 times out of 20 the true population mean is within two standard errors of the sample mean.

$\therefore$ 19 times out of 20 we can say the true population mean is between $454\frac{1}{2} \pm 2 \times 2.46 \simeq 449.6$ and 459.4 kilometres.

6. Confidence levels. It has been frequently emphasised that our conclusion will be correct 19 times out of 20. This of course is because 95 per cent (nineteen-twentieths) of all the means fall within two standard errors of the population mean. If the chance of being wrong 1 time out of 20 is too great a risk to take, we can be safer still by widening the range. If we extend it to three standard errors, then since 99¾ per cent of all means fall within three standard errors of the population mean we can be sure of being correct 399 times in 400 (99¾ per cent).

These different levels of certainty are known as *confidence levels*. Any estimate of a population mean must always indicate what level of confidence has been adopted.

7. Confidence limits and interval. In 6 above we saw how changes of confidence level led to changes in the distance between the mean limits. The limits are called the *confidence limits* since they are the limits determined by the chosen level of confidence. Logically, too, the interval between these limits is called the *confidence interval*. In the example given above, the 95 per cent confidence limits are 449.6 and 459.4 and the 95 per cent confidence interval is 459.4 − 449.6 = 9.8 kilometres.

Note that the higher the confidence level, the greater the confidence interval.

8. Summary: estimating a population mean.

(*a*) Take a random sample of n items.

(*b*) Compute the sample mean ($\bar{x}$) and the standard deviation (σ_x).

(*c*) Compute the standard error of the mean ($\sigma_{\bar{x}}$) from the formula $\sigma_{\bar{x}} = \sigma_x / \sqrt{n}$.

(*d*) Choose a confidence level (e.g. 95 per cent).

(*e*) Estimate the population mean as $\bar{x} \pm$ *appropriate number of* σ_x's (e.g. $\bar{x} \pm 2\sigma_{\bar{x}}$).

9. Effect of population size. It should be noted that nothing has been said above about the size of the population. This means that the *accuracy of our estimate is quite independent of population size*.

In other words, contrary to what seems "common sense," a population of 1 000 000 calls for no bigger sample than a population of 10 000. Our accuracy depends solely on sample size and the variability of the characteristic measured.

> NOTE: Strictly speaking, this is only true if the sample size is an insignificant proportion of the population size. If it is not an insignificant proportion, the real accuracy is actually *greater* than that claimed by the theory.

ESTIMATING POPULATION PROPORTIONS

10. Use of proportions. There are occasions in statistics when information cannot be given as a measure (e.g. kilometres, tonnes, minutes, pence, examination marks) but only as a *proportion*, such as males in a group of people, left-wing voters in an electorate, or defective production in total production. In these cases we are faced with estimating the population proportion from a single sample.

11. Sampling distribution of a proportion. If we again took a large number of samples of (say) 100 people from a particular population, we should not always find that we had the same number of males in every sample. If the population contained slightly more females than males, we might find that our samples contained anything from 30 to 60 males, i.e. proportions ranging between 0.3 and 0.6. A histogram of these proportions would result in another thin but definite normal curve of distribution having a mean equal to the true population proportion (*see* Fig. 35).

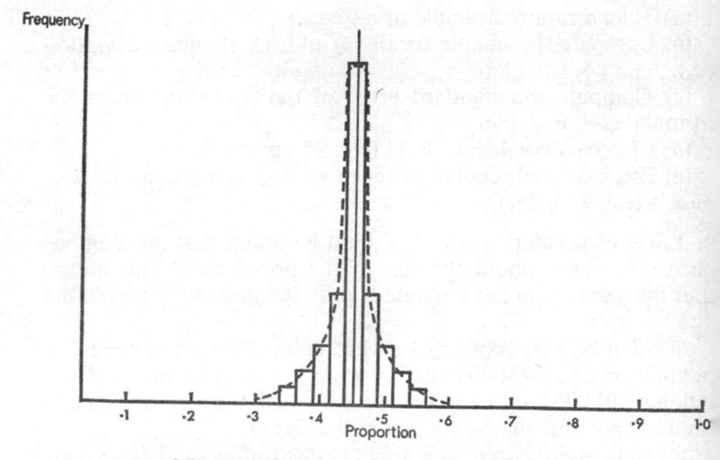

FIG. 35 *Sampling distribution of a proportion.*

This graph assumes the population proportion to be 0.45. It shows the frequency curve derived from the histogram of the various proportions found in a large number of samples.

12. Standard error of a proportion. Now, we can estimate the population proportion from our sample by using the sample proportion, but unless the standard deviation of this distribution —the *standard error of a proportion*—can be found, there is no way of determining the possible error in our estimate. Our other standard error formula is of no value here, since it is not possible to determine the standard deviation of a proportion from a single sample. How, for instance, could you calculate the standard deviation of a sample where 43 out of 100 people were male? Fortunately, there is yet another formula available which states that:

$$\textit{Standard error of a proportion } (\sigma_p) = \sqrt{\frac{pq}{n}}$$

where p = population proportion; $q = 1 - p$, and n = sample size.

Unfortunately this formula calls for the population proportion —the very thing we do not know. Substitution of the *sample* proportion for the population proportion is, however, usually allowed if the sample size is above 50.

13. Estimating a proportion. The procedure for estimating a proportion is similar to that for estimating a mean. If we know the standard error of a proportion, all we need do is compute:

$$\textit{Sample proportion} \pm 2\sigma_p$$

which will, of course, give us an estimate of the population proportion at the 95 per cent confidence level.

EXAMPLE

In a sample of 400 people, 172 were males. Estimate the population proportion at the 95 per cent confidence level.

$$\text{Sample proportion} = \frac{172}{400} = 0.43$$

$$\sigma_p = \sqrt{\frac{pq}{n}} = \sqrt{\frac{0.43 \times (1 - 0.43)}{400}} = \sqrt{\frac{0.43 \times 0.57}{400}}$$

$$= 0.0248$$

$\therefore$ Estimate of population proportion at the 95 per cent confidence level = sample proportion $\pm\ 2\sigma_p$.

$$= 0.43 \pm 2 \times 0.0248$$

$$= 0.3804 \text{ and } 0.4796$$

$$\triangleq \underline{\underline{\text{between 38 per cent and 48 per cent.}}}$$

14. Summary: estimating a population proportion.

(*a*) Take a random sample of n items, n being 50 or more.

(*b*) Compute the sample proportion.

(*c*) Compute the standard error of a proportion (σ_p) from formula $\sigma_p = \sqrt{(pq/n)}$, using the sample proportion for p.

(*d*) Choose a confidence level (e.g. 95 per cent).

(*e*) Estimate the population proportion as *Sample proportion* $\pm$ *Appropriate number of* σ_p's (e.g. Sample proportion $\pm 2\sigma_p$).

PROGRESS TEST 20

(*Answers in Appendix VI*)

1. Estimate the population mean (*a*) at the 95 per cent and (*b*) at the 99.75 per cent confidence levels, where the sample data is:

(*i*) Mean, 950 kg; σ, 15 kg; sample size, 25.
(*ii*) Mean, 1.82 cm; σ, 0.8 cm; sample size, 100.
(*iii*) Mean, 1.82 cm; σ, 0.8 cm; sample size, 10 000.

2. Estimate, at the 95 per cent level of confidence, the population proportion where sample data is:

(*i*) 61 males out of 100 people.
(*ii*) 6100 males out of 10 000 people.
(*iii*) 26 defectives out of 49 parts.

(*Warning: care is needed with the next two questions.*)

3. The result of a sample survey of 100 flowers of a particular type showed that the estimated mean flower height was 15 cm $\pm$ 2 cm at the 95 per cent level of confidence. The investigator decides that he needs an estimate which is within $\frac{1}{2}$ cm of the true population mean at this level of confidence. What must his sample size be?

4. If you wished to have a confidence level of 99 per cent in any survey, what would $\sigma_{\bar{x}}$ need to be multiplied by?

CHAPTER XXI

Tests of Significance

It frequently happens in statistical work that some fact is believed to be true, yet when a random sample is taken it turns out that the sample data does not wholly support the fact. The difference could be due to (*a*) the original belief being wrong, or to (*b*) the sample being slightly one-sided—as virtually all samples are to some degree.

Clearly, tests are needed to distinguish which is the more likely possibility. Such tests will reveal whether or not the difference could reasonably be ascribed to ordinary chance factors operating at the time the sample was selected. If the difference *cannot* be explained as being probably due solely to chance, the difference is said to be *statistically significant*. Tests devised to check whether this is so are called *significance tests*.

TESTING A HYPOTHESIS

The first of these tests aims to find out whether or not a belief about the mean of a population can continue to be held in the face of a sample that has a mean different from the believed population mean.

1. The null hypothesis. The approach here is to make the hypothesis, or assumption, that *there is no contradiction between the believed mean and the sample mean*, and that the difference can therefore be ascribed solely to chance. This hypothesis is called the *null hypothesis*, and the object of the test is to see whether the null hypothesis should be rejected or not.

2. Testing the null hypothesis. To do this, we make use of the sampling distributions of the means (i.e. the thin normal curve we obtained as a result of graphing the means of a large number of samples, in XX, 2). If the population mean is in fact the figure we believe it to be, then 95 per cent of the means of all samples will fall within two standard errors of this figure. All that is necessary, then, is to find out whether or not the sample mean does lie this close to the believed figure. If it does not, the true population mean cannot be the figure it is claimed to be—unless one takes the view that the sample mean just happens to be the one mean in twenty that lies outside the "two standard errors" limits.

3. Rejection of the null hypothesis. In other words if the sample mean lies more than two standard errors from the believed mean, one can reject the null hypothesis at the 95 per cent level of confidence and assert that since there *is* a contradiction between the believed population mean and the sample mean which cannot be explained by chance, then the population mean cannot be the figure it was believed to be.

EXAMPLE

Assume that Table VIII relates to the distance travelled by 120 salesmen taken at random from a very much larger field force. Someone now asserts that the mean distance travelled by all the salesmen in the field force is 460 kilometres. Our sample mean, however, is 454½ kilometres. Can the assertion be maintained at the 95 per cent level of confidence?

NOTE: Table VIII: $\sigma = 27$ km; $n = 120$.

(*a*) State the null hypothesis: "there is no contradiction between the asserted mean of 460 and the sample mean of 454½."

(*b*) Find the standard error of the mean. According to the standard error formula:

$$\sigma_{\bar{x}} = \frac{\sigma_x}{\sqrt{n}} = \frac{27}{\sqrt{120}} = 2.46 \text{ km}$$

(*c*) Compute limits within which the sample mean will fall 95 per cent of the time if the asserted mean is correct, i.e. asserted mean $\pm\, 2 \times \sigma_{\bar{x}}$

$= 460 \pm 2 \times 2.46 = 455.08$ and 464.92 kilometres.

(*d*) Check to see whether the sample mean does lie within those limits or not. In this case our sample mean of 454½ does *not* lie within them. Therefore the difference between the asserted mean (460) and the sample mean (454½) is *significant*, so the null hypothesis is rejected. Consequently, the assertion that the true mean is 460 kilometres cannot be held at the 95 per cent level of confidence.

4. Non-rejection of the null hypothesis. If the null hypothesis is rejected, we conclude that the population mean is not the figure originally asserted. But if the null hypothesis is *not* rejected it is important to appreciate that we do *not* conclude that the population mean *is* the figure asserted.

Non-rejection of the null hypothesis only signifies that there is no evidence that the true mean is not as asserted; it does not mean there is evidence that it is correct.

Thus if, in the above example, our sample mean had been 456, we would not have rejected the null hypothesis, although obviously the true population mean could as well be (say) 455 or 458—or virtually any figure in the 450's—as the asserted figure of 460. Testing a hypothesis cannot result in proof that the asserted figure is true. It may only show that such a figure is probably false.

5. Confidence level and the risk of rejecting a true hypothesis. In **3** we came to the conclusion that the population mean could not be 460 kilometres, unless our sample was the exceptional one in twenty. But what if the sample *was* the odd one in twenty? In that case we would have rejected the null hypothesis when it might be true and when 460 kilometres could in fact have been the true population mean.

The confidence level indicates the risk one takes of rejecting a null hypothesis that might well be correct. To dismiss the assertion that 460 kilometres was the population mean might lead to action being taken that could result in serious difficulties should the true mean turn out to be 460 after all. Under such circumstances it may be considered that the risk of being wrong one in twenty times is too great and that only a risk of one in four hundred, at most, is justifiable. In that case a confidence level of $99\frac{3}{4}$ per cent would be selected, i.e. the standard error would be multiplied by 3 when computing the confidence limits.

6. Confidence level and the risk of not rejecting an incorrect hypothesis. However, raising the confidence level can result in the opposite error, since it means that one risks clinging to a believed mean of 460 when in fact it is incorrect. In the example above, use of the $99\frac{3}{4}$ per cent confidence level brings the lower limit of tolerance down to $460 - 3 \times 2.46 = 452.62$. Since our sample mean is above this ($454\frac{1}{2}$) we would not reject the null hypothesis.

In short, the selection of the confidence level depends on which error is considered the graver: to reject a hypothesis which may be true, or to fail to reject one that is wrong.

Finally, note that it is essential to select the confidence level *before* testing the hypothesis. Errors of interpretation could arise if one measured the deviation of the sample mean from the asserted population mean first, and then computed the chance of such a difference arising.

7. Type I and II errors. Sections **5** and **6** indicate that whenever we make a test of significance, there are two diametrically opposite types of errors into which we can fall—that of rejecting the null hypothesis when it is true (e.g. the asserted mean is true but we say it is not), or that of not rejecting the null hypothesis when it is false (e.g. the asserted mean is false but we say there are no grounds for disbelieving it). These errors are known as type I and type II errors respectively and can be defined more generally as follows:

A *type I error* is the *error of rejecting a hypothesis when it is in fact true.*

A *type II error* is the *error of not rejecting a hypothesis when it is in fact false.*

Clearly, we can only decrease the risk of one type of error by increasing the risk of the other. Thus, as we saw in **5** and **6**, increasing the confidence level results in the reduction of the probability of making a type I error but increases the probability of making a type II error. In practice, then, before starting a significance test we must decide on the importance of each type of error and then set the confidence level accordingly. For example, an assertion that a given level of radio-activity was safe would be tested at a very high level of confidence since refusing to accept the assertion even if it were true would be far less disastrous than accepting it, should it prove to be false.

8. One-tail and two-tail tests. At this juncture a rather subtle point arises. In significance testing one should be quite sure just what is being tested. Can you, for instance, see the difference between the two following assertions:

(*a*) "The average weight of meat in 50 kg of my sausages is 35 kg."

(*b*) "The average weight of meat in 50 kg of my sausages is not less than 35 kg."

Clearly, in (*a*) what is being asserted is that the average is 35 kg—neither more nor less—while what is being asserted in (*b*) is that the average weight is not less than 35 kg, *though it could be more*. Now if the average weight *was* 35 kg, the standard deviation was, say, 2 kg and 50 kg samples having sample sizes of 64 were taken, then a distribution of the means would be obtained that had a mean of 35 kg and a standard error of $\frac{2}{\sqrt{64}} = 0.25$ kg. So 95 per cent of all samples will have means that lie inside the range $35 \pm 2 \times 0.25 = 34.5$ to 35.5 kg—and

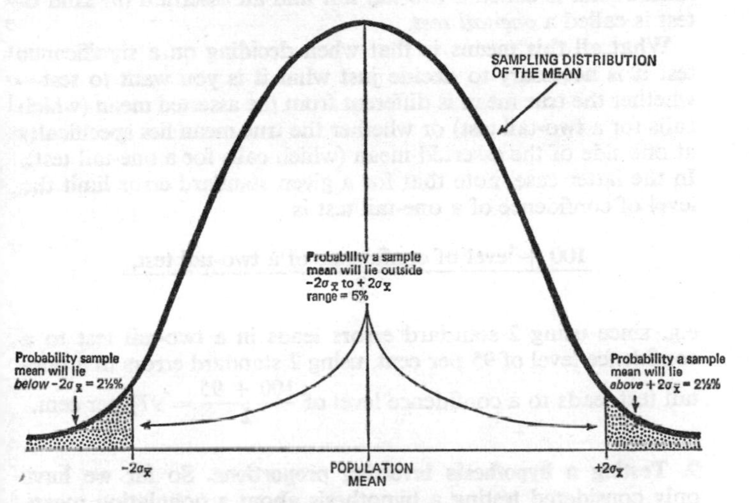

FIG. 36 *One-tail and two-tail probabilities.*

5 per cent will lie *outside* this range. So there will be only 1 chance in 20 of a sample having a mean weight below 34.5 kg or above 35.5 kg. But note, *this 5 per cent is made up of that area under the tail of a normal curve that lies below 34.5 kg and that area that lies above 35.5 kg* (*see* Fig. 36)—i.e. it involves *two tails* of the curve. However, if we only want the probability of taking a sample the mean of which lies *below* 34.5 kg then we must only measure the area below the lower *one tail* of the curve—and this, obviously, is half 5 per cent, i.e. 2½ per cent.

Now let us apply this to significance tests. If assertion (*a*) above is to be tested, then we will be equally interested in a population mean that is above 35 kg as one that is below. If our confidence interval is 34.5 to 35.5 kg, we would be testing the assertion at the 95 per cent level of confidence since 95 per cent sample means would lie within this range. If, however, assertion (*b*) is to be tested, we would only be interested in a population mean that fell *below* 35 kg, and if our confidence limit was 34.5 kg, we would be testing the assertion as the 97½ per cent *level of confidence* since only 2½ per cent of sample means would lie below this limit (*see* Fig. 36 again). For obvious reasons an assertion (*a*) kind of test is called a ***two-tail test*** and an assertion (*b*) kind of test is called a ***one-tail test***.

What all this means is that when deciding on a significance test it is necessary to decide just what it is you want to test—whether the true mean is different from the asserted mean (which calls for a two-tail test) or whether the true mean lies specifically at one side of the asserted mean (which calls for a one-tail test). In the latter case, note that for a given standard error limit the level of confidence of a one-tail test is

$$\frac{100 + \text{level of confidence of a two-tail test,}}{2}$$

e.g. since using 2 standard errors leads in a two-tail test to a confidence level of 95 per cent, using 2 standard errors in a one-tail test leads to a confidence level of $\frac{100 + 95}{2} = 97\frac{1}{2}$ per cent.

9. Testing a hypothesis involving proportions. So far we have only considered testing a hypothesis about a population mean. If now we wish to test a hypothesis about a *population proportion* (e.g. we wish to test an assertion that the proportion of Labour voters in a constituency is 60 per cent) then we simply have to substitute the word "proportion" for "mean" in all the foregoing paragraphs.

NOTE: Since the standard error of a proportion is equal to $\sqrt{(pq/n)}$ where ideally p is the population proportion (*see* XX, 12) then the *asserted* proportion must be used in calculating the standard error of the proportion. This is because we must accept the asserted proportion as the population proportion until *after* we have at least proved it incorrect.

10. Summary: testing a hypothesis. The object is to test whether the difference between an asserted mean (proportion) and a sample mean (proportion) is significant or not.

(*a*) State the null hypothesis.

(*b*) Select the confidence level required on the basis of the importance of a type I and a type II error.

(*c*) Compute the standard error of the mean (proportion) from the appropriate standard error formula.

(*d*) Compute *asserted mean* $\pm y \times \sigma_{\bar{x}}$ (or *asserted proportion* $+ y \times \sigma_p$) where y is the appropriate factor (taken from tables) in view of the level of confidence selected and bearing in mind whether a one-tail or a two-tail test is called for.

(*e*) Check where the sample mean (proportion) falls:

(*i*) If *outside* these limits, reject the null hypothesis, i.e. the population mean (proportion) is not as asserted.

(*ii*) If *inside* these limits, do not reject the null hypothesis, i.e. the population mean (proportion) could be as asserted.

TESTING DIFFERENCES BETWEEN MEANS AND PROPORTIONS

11. Difference between means. It is quite common in statistical work to be confronted with two distinct populations which seem likely to have virtually identical means. For example, cats in the north of England are probably the same height as those in the south, that is, the mean heights of cats in the two populations are the same. Nevertheless, if a sample were taken from each population it would be unlikely that the two sample means would be identical. How could we tell whether the difference between the sample means was due solely to chance factors, or to a real difference between the two population means? In other words, how can we tell whether the difference between the means is significant?

12. Distribution of the difference between means. Assume that the two population means were the same. In that case, if a great many pairs of samples were taken and the difference found between the means of each pair (always deducting the mean heights of the southern cats from the mean heights of the northern cats), the differences would be found to be small—indeed, there would be occasions when there was no difference. When there were differences, about half of them would be plus and half minus.

On only a very few occasions would there be large differences, plus or minus.

If these differences were graphed it would be found that once again the distribution would follow a normal curve, this time one with a mean of zero and extending over plus and minus values (*see* Fig. 37). And, again, 95 per cent of the differences would lie within two standard errors of the mean of zero; in other words, *95 per cent of the differences would not exceed two standard errors.*

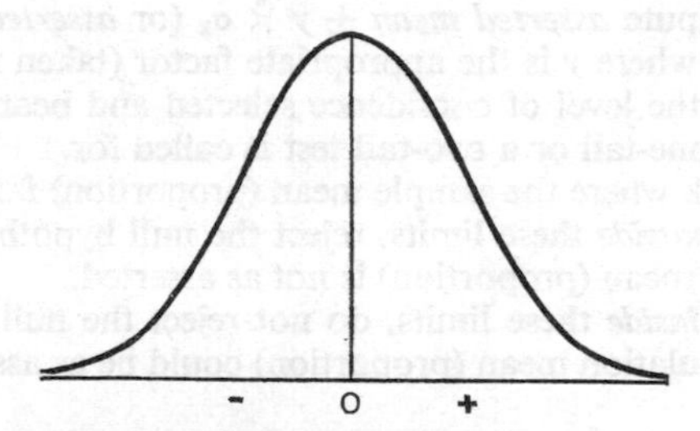

FIG. 37 *Sampling distribution of the difference between means.*

If a large number of pairs of samples were taken from two populations having identical means—one sample from each population—and the difference between each pair of sample means found, then a graph of the differences would result in the normal curve shown here.

13. Testing a difference between means. The original problem now becomes one of testing to see whether or not an actual difference found exceeds two standard errors in a distribution of differences between means. If it does not exceed two standard errors, the difference can be set down to chance and one can conclude there is no evidence to prove that the two populations do not have the same mean.

First, however, it is necessary to find the standard error in a distribution of differences between means. This figure is given by the formula:

$$\sigma_{(\bar{x}_1 - \bar{x}_2)} = \sqrt{\sigma^2_{\bar{x}_1} + \sigma^2_{\bar{x}_2}}$$

where $\sigma_{(\bar{x}_1 - \bar{x}_2)}$ = the standard error of the difference between means.

We check, therefore, to see whether or not the actual difference exceeds $2\sigma_{(\bar{x}_1-\bar{x}_2)}$. If it does, the population means are most unlikely to be the same. In statistical jargon we should say that "the difference between the means is significant at the 95 per cent level of confidence, and therefore the populations do not have the same mean." Of course, other confidence levels could be selected, in which case $\sigma_{(\bar{x}_1-\bar{x}_2)}$ would need to be multiplied by the appropriate factor.

EXAMPLE

Assume that:

North of England cats (x_1): sample mean 25 cm; standard deviation of sample $6\frac{1}{4}$ cm; sample size 100.
South of England cats (x_2): sample mean 24 cm; standard deviation of sample 6 cm; sample size 144.

Now $\sigma_{\bar{x}} = \dfrac{\sigma_x}{\sqrt{n}}$

$$\therefore \sigma_{\bar{x}_1}, = \frac{6.25}{\sqrt{100}} = 0.625 \text{ cm}$$

and

$$\sigma_{\bar{x}_2}, = \frac{6}{\sqrt{144}} = 0.5 \text{ cm}$$

And using formula above:

$$\sigma_{\bar{x}_1,-\bar{x}_2} = \sqrt{\sigma^2_{\bar{x}_1} + \sigma^2_{\bar{x}_2}} = \sqrt{0.625^2 + 0.5^2} = 0.8 \text{ cm}$$

Now, if the 95 per cent level of confidence is required (and note that this is a two-tail test), limits are

$$2\sigma_{\bar{x}_1-\bar{x}_2} = 2 \times 0.8 = 1.6 \text{ cm.}$$

And the actual difference between means $= 25 - 24 = 1$ cm. Since this 1 cm is within the 1.6 cm limit, the difference could have arisen through chance factors and is not significant at the 95 per cent level of confidence. There is no evidence, therefore, that the cats in the two parts of England have different heights.

14. One-tail and two-tail tests. Again note we should really first consider if we are to use a one-tail or two-tail test. A glance at Fig. 36 will indicate that whereas 5 per cent of the differences will lie *outside* the two-standard-errors-from-the-mean range if there are no population mean differences, only $2\frac{1}{2}$ per cent will lie *beyond* a given two-standard-error limit. In other words, in our example above we tested at the 95 per cent level of confidence whether or not there was a significant difference between the heights of cats in the north of England and the height of cats in the south. If we had been testing whether northern cats were *higher* than southern cats, then a one-tail test would be involved and that means from our results we could say that the difference between the means was not significant at the 97½ per cent level of confidence. (In this particular case, of course, it is obvious that if there is no significance at the 95 per cent level of confidence there certainly cannot be any at a higher level. However, where tests reveal significance this aspect becomes important.)

15. Summary: testing a difference between means. The object is to test whether or not the difference between two sample means is significant.

(*a*) Select the level of confidence required on the usual basis.

(*b*) Find the standard error of the mean ($\sigma_{\bar{x}}$) for both samples.

(*c*) Compute the standard error of the difference between means by formula $\sigma_{(\bar{x}_1-\bar{x}_2)} = \sqrt{\sigma^2{}_{\bar{x}_1} + \sigma^2{}_{\bar{x}_2}}$

NOTE: Since $\sigma_{\bar{x}} = \dfrac{\sigma_x}{\sqrt{n}}$, this standard error can be calculated in one step by using the formula:

$$\sigma_{(\bar{x}_1-\bar{x}_2)} = \sqrt{\frac{\sigma^2{}_{\bar{x}_1}}{n_1} + \frac{\sigma^2{}_{\bar{x}_2}}{n_2}},$$

(*d*) Multiply $\sigma_{(\bar{x}_1-\bar{x}_2)}$, by the appropriate factor for the level of confidence selected.

(*e*) Find the actual difference between the two sample means. If the difference is below the limit found in step (*d*), it is not significant. If it is above that limit, the difference is significant and the conclusion can be drawn (at the chosen level of confidence) that the two populations have different means.

16. Testing a difference between proportions. Problems sometimes involve testing a difference between *proportions* in samples instead of means. This gives little trouble, since the procedure outlined above can still be applied except that the word "proportion" should be substituted for the word "mean." Even the formula for the standard error of the difference between proportions is the same, although of course the symbols change slightly, i.e.:

$$\sigma_{(p_1-p_2)} = \sqrt{\sigma^2{}_{\text{prop}_1} + \sigma^2{}_{\text{prop}_2}}$$

NOTE: Since $\sigma^2{}_p = \left(\sqrt{\dfrac{pq}{n}}\right)^2 = \dfrac{pq}{n}$

then the formula can be written as:

$$\sigma_{(p_1-p_2)} = \sqrt{\frac{p_1q_1}{n_1}+\frac{p_2q_2}{n_2}}$$

NOTE: There is a refinement of this formula. As we saw in **12**, the distribution of differences, on which the test is based, is the distribution we obtain when, in fact, there are no *population* differences. So a standard error of a difference between two proportions is computed in respect of a situation where there is in effect only one population. This being so, it would improve our formula for $\sigma_{(p_1-p_2)}$ if a more accurate estimate of the overall population proportion were employed (remember, all standard errors relating to proportions should, strictly speaking, be based on the population proportions, not the sample proportions). If, then, the two samples are initially assumed to come from a single population we can make such an estimate by "pooling" the two sample results—i.e. by adding together the numbers having the characteristic concerned in the two samples and dividing by the combined samples total. For instance, if 210 people out of a first sample of 400 wore glasses and 320 out of a second sample of 500, we could pool these figures and say that our best estimate of the population proportion was $\frac{210+320}{400+500}=\frac{530}{900}=0.59$. This proportion would then be used in our $\sigma_{(p_1-p_2)}$ formula in lieu of p_1 and p_2.

To summarise, our formula can be improved a little by writing it as:

$$\sigma_{(p_1-p_2)} = \sqrt{\frac{pq}{n_1}+\frac{pq}{n_2}} = \sqrt{pq\left(\frac{1}{n_1}+\frac{1}{n_2}\right)},$$

where p = pooled proportion and $q = 1 - p$.

PROGRESS TEST 21

(*Answers in Appendix VI*)

1. A child welfare officer asserts that the mean sleep of young babies is 14 hours a day. A random sample of 64 babies shows that their mean sleep was only 13 hours 20 minutes, with a standard deviation of 3 hours. Test the officer's assertion at the 95 per cent level of confidence.

2. An election candidate claims that 60 per cent of the voters support him. A random sample of 2500 voters show that 1410 support him. Test his claim at the $99\frac{3}{4}$ per cent level of confidence.

3. A sample of 200 fish of a particular species taken at random from one end of a lake had a mean weight of 20 kg and a standard deviation of 2 kg. At the other end of the lake, a sample of 80 fish showed a mean weight of $20\frac{1}{2}$ kg and a standard deviation of 2 kg also. An expert on fish claimed that these fish swam all over the lake and the two samples were therefore taken, in effect, from the same population. Test this assertion at the 95 per cent level of confidence.

4. A health official claims that the citizens of city A are fitter than those of city B, and in evidence shows that 96 out of 200 citizens of city A, selected at random, passed a standard fitness test as against only 84 out of 200 citizens of B. Do you think he has proved his claim?

CHAPTER XXII

Sampling with Small Samples

In the previous chapters it was emphasised that the theory outlined applied only to large samples. In this chapter the theory as applied to small samples is examined.

THE *t* DISTRIBUTION

1. From large to small samples. In XX, 2 and 3, we said that the distribution of the means of thousands of samples of a given size taken from a particular population would be normal and where large samples were concerned the standard error of such a distribution ($\sigma_{\bar{x}}$) could be estimated from the formula $\sigma_{\bar{x}} = \frac{\sigma_{\text{sample}}}{\sqrt{n}}$. Using this formula and a normal curve table it was, therefore, possible to estimate the population mean at any desired level of confidence.

Now unfortunately, with increasingly smaller samples this increasingly fails to be true. There are, in fact, two adjustments that are needed (and these are explained in **2** and **3** below), though once these have been made the procedure for estimating a population mean from a small sample is exactly the same as that of estimating a population mean from a large sample.

2. Formula for estimating a population standard deviation. The first adjustment relates to σ_{sample} in our formula $\sigma_{\bar{x}} = \frac{\sigma_{\text{sample}}}{\sqrt{n}}$. Remember, as we said in XX, 3, strictly speaking the formula should be $\sigma_{\bar{x}} = \frac{\sigma_{\text{population}}}{\sqrt{n}}$, but if the standard deviation of the population is unknown (as it generally is) we can use σ_{sample} as an estimate in lieu. However, using σ_{sample} as an estimate of $\sigma_{\text{population}}$ leads to a very slight bias since usually a sample standard deviation is a little smaller than the population standard deviation. With large samples the bias is trivial but it increases with small samples and so must be allowed for. Fortunately this is simple enough—instead of dividing by n in the formula $\sigma = \sqrt{\frac{\Sigma(x - \bar{x})^2}{n}}$ one uses $n - 1$. In other words:

$$\text{Estimate of population standard deviation} = \sqrt{\frac{\Sigma(x - \bar{x})^2}{n - 1}},$$

where x, $\bar{x}$ and n all relate to the sample.

NOTES

(*i*) Beware of confusion here. The standard deviation of a sample is always $\sqrt{\frac{\Sigma(x - \bar{x})^2}{n}}$. However, if we want an *estimate of the population standard deviation* we use in lieu $\sqrt{\frac{\Sigma(x - \bar{x})^2}{n - 1}}$. The $n - 1$ adjusts for the bias in making an estimate of the population standard deviation—it does nothing to improve (in fact, it is erroneous and irrelevant to) the calculation of the *sample* standard deviation.

(*ii*) The insignificance of the bias in the case of large samples can be judged from our formula above, since when n is large deducting 1 makes virtually no difference (e.g. in a sample of 50 using just n instead of $n - 1$ results in a mere 1 per cent error in the estimate).

(*iii*) Do not mix up the "n" in this formula with the "n" in the $\sigma_{\bar{x}} = \sigma_{\text{population}}/\sqrt{n}$ (*see* 3).

3. Accuracy of estimate of population standard deviation. Regrettably, the adjustment above is not the end of our difficulties as regards the population standard deviation. As can be imagined, with a very small sample of, say, three or four items the standard deviation of such a sample would depend to some extent on chance—the items could fortuitously be closely grouped or widely spaced in relation to the population distribution—and so we can be less sure of the accuracy of our estimate of the population standard deviation. Since we obtain the standard error of the mean from the estimate of the population mean (i.e. $\sigma_{\bar{x}} =$ Estimate of population standard deviation$/\sqrt{n}$), it follows that we can be less confident that the population mean lies within, say, 2 standard errors of the sample mean when taking small samples.

This situation has been studied and as a result it has been found that we can still use our ordinary sampling procedure (with the minor adjustment given in **2**, of course) provided that instead of using a normal curve for setting our confidence limits we use what is called a *t distribution* curve. In Fig. 38 such a curve is compared with a normal curve. As can be seen, the *t* distribution curve is a little flatter than a normal curve and this means that one must go further out from the centre in order to reach the two points that between them enclose any given area under the curve—i.e. the confidence interval needs to be wider for a given level of confidence.

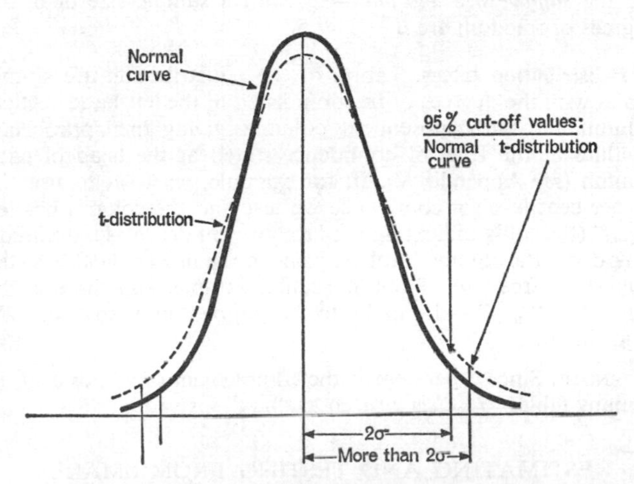

FIG. 38 *Comparison of a normal distribution and a typical t distribution.*

The normal curve is shown by the continuous line and the *t* distribution by the dotted line. The difference between the curves is exaggerated a little for the sake of clarity.

4. Sample size and the *t* distribution. A feature of the *t* distribution is that it changes with sample size, being relatively flat for a sample size of 2 and gradually becoming more and more like a normal curve as the sample size increases. This means that the confidence limits associated with any given level of confidence will change with the sample size.

Now in order to avoid publishing a separate table for the *t* distribution areas for every sample size, a single table is published which merely lists the areas relating to a selected number of levels of confidence.

5. Degrees of freedom. One minor complication at this point arises from the need in more advanced statistics to use what are called *degrees of freedom.* At this level, understanding the concept is not vital—the only trouble is that published tables refer to degrees of freedom and not sample size. However, in this context the conversion is easy since the degrees of freedom are the *sample size less one*—e.g. with a sample size of 6, the degrees of freedom are $6 - 1 = 5$.

6. *t* distribution tables. Tables of the *t* distribution are simply tables with the degrees of freedom listed in the left-hand vertical column and with subsequent columns giving the appropriate *t* values at the level of confidence stated at the head of each column (*see* Appendix V). If, for example, we wish to use the 95 per cent level of confidence we just find the column headed "$t_{0 \cdot 95}$" (the suffix indicating the level of confidence as a decimal), look down the column until we come to the line applicable to the degrees of freedom of our particular sample, and there is the value for "$t_{0 \cdot 95}$"—all ready to be slipped into any formula requiring it.

NOTE: Since 5 per cent is the complement of 95 per cent, in many tables "$t_{0 \cdot 95}$" is written as "$t_{0 \cdot 05}$" instead.

ESTIMATING AND TESTING FROM SMALL SAMPLES

7. Estimating a population mean from a small sample. When we were taking large samples we found that we could estimate the population mean at the 95 per cent level of confidence by using the following formula:

$$\text{Population mean} = \text{sample mean} \pm 2\sigma_{\bar{x}}.$$

Taking small samples, the method is *exactly the same except that we must multiply the standard error by* $t_{0 \cdot 95}$, where $t_{0 \cdot 95}$ is the value given by the *t* distribution table for the appropriate degrees of freedom involved. So the estimate of the population mean from a small sample at the 95 per cent level of confidence is:

$$\text{Population mean} = \text{Sample mean} \pm t_{0 \cdot 95} \times \sigma_{\bar{x}}.$$

Apart from the adjustment detailed in **2**, then, the only difference is that the figure we multiply the standard error by is taken from a t distribution table instead of a normal curve table.

8. Summary—estimation of population mean from a small sample. In summary, to estimate a population mean from a sample of size below 30:

(*a*) Find the sample mean.

(*b*) Estimate the population standard deviation from the formula

$$\text{Estimate } \sigma_{\text{population}} = \sqrt{\frac{\Sigma(x - \bar{x})^2}{n - 1}}$$

(*c*) Find the standard error of the mean from the formula $\sigma_{\bar{x}} = \text{estimate of } \sigma_{\text{population}}/\sqrt{n}$

(*d*) Find the degrees of freedom (d. of f.) applicable to sample (d. of f. $= n - 1$)

(*e*) Decide on the level of confidence required and use the t distribution table to determine the required value of t at the chosen level of confidence and relevant degrees of freedom.

(*f*) Apply the formula:

Estimate of population mean = sample mean $\pm\, t \times \sigma_{\bar{x}}$

EXAMPLE

A political scientist, wishing to see if there is anything to support a theory that the political interests of a person are related to his height, decides to take a random sample of U.K. councillors and find their heights. However, to keep costs within bounds he limits his sample to 10—the selected councillors proving to have heights (in feet and decimal of a foot) of 5.5, 6.1, 5.4, 5.8, 5.3, 4.8, 5.2, 4.6, 6.1 and 5.2. Estimate the mean height of all councillors at the 95 per cent level of confidence.

SOLUTION

(*a*) $\Sigma x = 5.5 + 6.1 + 5.4 + 5.8 + 5.3 + 4.8 + 5.2 + 4.6 + 6.1 + 5.2 = 54.0$

$$\therefore\ \bar{x} = \frac{\Sigma x}{n} = \frac{54}{10} = 5.4 \text{ ft}$$

(*b*) $\Sigma(x - \bar{x})^2 = 0.01 + 0.49 + 0 + 0.16 + 0.01 + 0.36 + 0.04 + 0.64 + 0.49 + 0.04 = 2.24$

$$\therefore \text{ Estimate population } \sigma = \sqrt{\frac{\Sigma(x - \bar{x})^2}{n - 1}}$$

$$= \sqrt{\frac{2.24}{10 - 1}} = 0.50 \text{ ft}$$

(*c*) $\sigma_{\bar{x}} = \dfrac{\text{Estimate population } \sigma}{\sqrt{n}} = \dfrac{0.50}{\sqrt{10}} = 0.158 \text{ ft}$

(*d*) d. of f. $= n - 1 = 10 - 1 = 9$

(*e*) $t_{0 \cdot 95} = 2.262$*

(*f*) $\therefore$ Estimate of mean height of all councillors at the 95 per cent level of confidence $= \bar{x} \pm t_{0 \cdot 95} \times \sigma_{\bar{x}}$

$= 5.4 \pm 2.262 \times 0.158 = \underline{\underline{5.04 \text{ to } 5.76 \text{ ft}}}$

9. Testing a hypothesis. If a hypothesis involving a small sample is to be tested, then the large sample testing procedure can be used provided that it is adapted in line with the above theory—though note that in the case of a test of differences between means the relevant degrees of freedom is "sample size − 2" (i.e. $n - 2$). An example of such a test is given in **10** below.

10. Testing paired differences. A useful application of a *t* distribution test relates to where *pairs* of experiments are made such that one half of each pair is influenced by a given factor and the other half left uninfluenced. For example, a new fertiliser could be tested by taking a number of different plots (to ensure the fertiliser was not effective in just one environment), each plot being uniform over its entire area, dividing these plots into halves and treating one half with the usual fertiliser and the other with the new. The yields are then measured and recorded in pairs.

Now at first sight it may appear that the most straightforward way of testing would be to find the mean yield of the halves treated with the new fertiliser and the mean yield of the other halves and test the difference between the means. However, the variability of the yields between *plots* would almost certainly lead to such large standard errors that it would be virtually impossible for the test to detect any significant differences. A better approach is to find the differences *between each pair* of yields, find the mean difference and then on the hypothesis that

* This estimate calls for the two-tail value.

the new fertiliser is no better than the usual one, so that all differences are due to chance and the population mean difference is really zero, test if the actual mean difference varies significantly from zero.

EXAMPLE

An educational psychologist suspects that I.Q. test A tends to give higher I.Q. scores than I.Q. test B. He gives both tests to five boys and obtains the following results:

Boy	*1st*	*2nd*	*3rd*	*4th*	*5th*
Test A	135	103	129	96	121
Test B	125	102	117	94	121

He claims the results confirm his suspicions. Test his claim at the 97½ per cent level of confidence.

SOLUTION

Boy	*1st*	*2nd*	*3rd*	*4th*	*5th*	Σ
Test A	135	103	129	96	121	
Test B	125	102	117	94	121	
Difference (x)	+10	+1	+12	+2	0	+25
$(x - \bar{x})^2$	25	16	49	9	25	124

$$\bar{x} = \frac{+25}{5} = +5$$

Estimated population standard deviation

$$= \sqrt{\frac{124}{5-1}} = 5.568.$$

$$\text{Standard error } (\sigma_{\bar{x}}) = \frac{5.568}{\sqrt{5}} = 2.490.$$

So the sample mean (i.e. 5 I.Q. marks) lies $\frac{5}{2.49} = 2.008$ standard errors from zero, the hypothetical population mean. Degrees of freedom $= 5 - 1 = 4$.

Now since this is a one-tail test (we are testing if A gives a *higher* score, not a different score), the t value for a 97½ per cent level of confidence is the same as the t value for a two-tail 95 per cent level of confidence test—and tables show this to be 2.776. As the deviation of the actual sample mean is less than this number of standard errors from zero, the differences are not significant at the 97½ per cent level of confidence. The psychologist's suspicions, therefore, remain unproved.

PROGRESS TEST 22

(Answers in Appendix VI)

1. A random sample of 16 components has a mean of 6.214 cm and a standard deviation of 0.120 cm. Estimate the population mean at the 95 per cent level of confidence.

2. A random sample of 3 apples in a barrel was taken and weighed, the weights proving to be 122, 133 and 102 g respectively. If the net weight of the apples in the barrel was 45.5 kg, estimate the number of apples in the barrel at the 95 per cent level of confidence.

3. A research worker engaged in testing if alcohol had any effect on the speed of a person's responses selected 6 volunteers. These he tested for speed of response to various stimulae, collectively measuring an individual's speed by a "speed factor." He then gave each volunteer a capsule which, unknown to the volunteer, contained a standard alcohol measure. He then repeated his speed of response tests. His results were as follows:

Volunteer	*A*	*B*	*C*	*D*	*E*
Speed Factor:					
Before	12.5	10.2	11.7	9.4	12.1
After	12.9	10.3	12.5	9.6	12.6

Test at the 95 per cent level of confidence if the speed of response after taking the capsule was significantly different from before.

CHAPTER XXIII

χ^2 *Tests*

Imagine that we threw a die 600 times. We would, of course, expect each face to fall upward 100 times. If, however, the "2" fell upwards 102 times we would not be surprised—we would simply explain the difference between 100 and 102 as one arising by chance. On the other hand, if the "2" fell upwards 300 times we would feel that this was extremely unlikely to be due to chance and that something was wrong with the die. The problem is, however, where do we draw the line? When must one stop saying "Ah, yes, the difference between actual and expected result can be explained by chance" and start saying "The difference is too big to be reasonably explained by chance"?

In this situation what we need is some sort of significance test.

THE χ^2 DISTRIBUTION

One such test involves what is called the χ^2 (pronounced *Ki*—as in "Kite"—*squared*) distribution. This distribution, which proves to be very useful in a variety of different circumstances, is used to *test if an observed series of values differs significantly from what was expected* ("significant" meaning "statistically significant"—*see* p. 165. In this section the nature of this distribution is examined while in the two subsequent sections applications of the distribution are illustrated.

1. The χ^2 formula. χ^2 is a numerical figure associated with a particular group of observations. The formula for finding this figure from a given group of observed actual values is:

$$\chi^2 = \sum \frac{(\text{Actual} - \text{Expected value})^2}{\text{Expected value}}$$

i.e. to find χ^2, square the difference between the actual and expected values of each item in the series, divide by the expected value, and then add up all the quotients.

NOTE: We never bother with the square root of χ^2.

2. Implications of the χ^2 formula. Let us consider what the χ^2 formula implies. First, if the actual results are exactly as expected then (Actual − Expected) = 0 for each item in the series, and χ^2 works out at zero. On the other hand, the bigger the differences between actuals and expected, the bigger the square of these differences and therefore the bigger χ^2 becomes.

NOTE: Dividing by the expected value simply brings the squared difference into proportion—e.g. a difference of 10 when 40 was expected is a very much "bigger" difference than one of 10 when 4000 was expected.

3. The χ^2 distribution. Next let us consider throwing a perfectly good die 600 times—and repeating this 600-throw cycle over and over again, finding χ^2 for each cycle. Now the number of cycles in which we would have each face falling upwards exactly 100 times (χ^2 therefore equalling 0) would be virtually none. More frequently a difference of between 5 and 10 from the expected value of 100 would arise (if, say, there were a difference of 5 for every face then

$$\chi^2 = \frac{5^2}{100} + \frac{5^2}{100} + \frac{5^2}{100} + \frac{5^2}{100} + \frac{5^2}{100} + \frac{5^2}{100} = 1{\cdot}5).$$

Very rarely differences of, say, 20 would arise (and if they did, again assuming they did for every face, then

$$\chi^2 = 20^2/100 \times 6 = 24).$$

Clearly, if we carried out this exercise enough times we could put all our χ^2 into a *distribution* and prepare a frequency curve. From the preceding discussion it will be appreciated that such a curve would probably look like the one in Fig. 39. Now obviously at some point towards the extreme right of the curve we can draw a line that cuts off the upper 5 per cent of the total area—e.g. we can find a value such that 95 per cent of the ascertained values of χ^2 will lie between this number and 0.

4. Using the χ^2 distribution. Our distribution curve above was formed as a result of using a perfectly good die. Assume we now use a die the perfection of which is in some doubt. We throw it 600 times and compute χ^2 from the results. What if our figure lies *above* the 95 per cent cut-off value on the distribution? We are then faced with the usual dilemma—i.e. is our result one of the 1 in 20 that can arise with a good die, or is our die not so good?

Clearly, we would only be wrong once in twenty times if we elected to say such a die was not a good one. In other words, we can use our χ^2 distribution to test whether a group of differences between actual and expected results is statistically significant.

5. The χ^2 cut-off values. Of course, to use the χ^2 distribution in this way we need to know what cut-off values to use. Now a moment's thought will show us that the χ^2 distribution will depend upon the number of categories in the series (e.g. in the example above there were 6 die faces involved and so 6 categories) since the more categories there are the larger χ^2 can be without its value being exceptional.

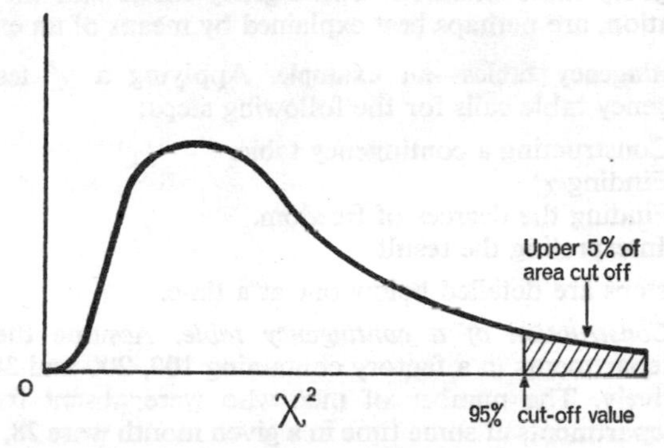

FIG. 39 *Typical χ^2 distribution.*

(NOTE: In the example discussed in 3 the χ^2 scale would be such that 1.5 would be on the far left of the axes while 24 would be off the page on the right-hand side.)

We need, therefore, tables that give us cut-off values that allow for confidence levels and numbers of categories. Now there is a link (though a rather complex one) between categories and degrees of freedom. As a result χ^2 tables are similar in layout to *t* distribution tables—i.e. degrees of freedom are listed in a vertical column and in the subsequent columns, headed with various levels of confidence, are the χ^2 cut-off values (*see* Appendix V). Remember, the χ^2 test is based on the fact that if the value of χ^2 computed from the observed data exceeds the cut-off values in the table, the differences between actual and expected results are unlikely (at your chosen level of confidence) to be due solely to chance.

NOTE: We rarely concern ourselves with other than the upper tail of the distribution so deciding whether the situation calls for a one-tail or a two-tail test does not often arise when making a χ^2 test.

Now that we know what the χ^2 distribution is, the next two sections will illustrate its application to significance testing.

CONTINGENCY TABLES

Probably the most common application of the χ^2 test involves a contingency table situation. Contingency tables and the χ^2 test application, are perhaps best explained by means of an example.

6. Contingency tables—an example. Applying a χ^2 test to a contingency table calls for the following steps:

(*a*) Constructing a contingency table.
(*b*) Finding χ^2.
(*c*) Finding the degrees of freedom.
(*d*) Interpreting the result.

These steps are detailed below one at a time.

(*a*) *Construction of a contingency table.* Assume there are three departments in a factory containing 100, 200 and 300 men respectively. The number of men who were absent from the three departments at some time in a given month were 28, 70 and 82 respectively. A contingency table illustrating this information could be constructed as follows:

	Absent	*Not absent*	*Total*
Department 1	28	72	100
,, 2	70	130	200
,, 3	82	218	300
Total	180	420	600

Given this table it might well be asked if there was any significant difference in absenteeism between the departments.

(*b*) *Finding* χ^2. If there were no significant difference between departments in respect of absenteeism then we would expect departmental absenteeism to be running at the overall average rate. Since a total of 180 out of 600 were absent this average rate is $180/600 \times 100 = 30$ per cent. In Department 1, therefore, we would expect 30 out of the 100 men to be absent, and absenteeism in Departments 2 and 3 to be 60 and 90 respectively. Let us now insert in brackets in *each and every* cell of our contingency table these *expected* results.

NOTE: Totals and row and column headings are not regarded as occupying contingency table cells.

	Absent	*Not absent*	*Total*
Department 1	28 (30)	72 (70)	100
,, 2	70 (60)	130 (140)	200
,, 3	82 (90)	218 (210)	300
Total	180	420	600

From this table we can find χ^2 as follows:

	Actual (*A*)	*Expected* (*E*)	(*A–E*)	$(A-E)^2$	$\frac{(A-E)^2}{E}$
Department 1:					
Absent	28	30	−2	4	0·1333
Not absent	72	70	+2	4	0·0571
Department 2:					
Absent	70	60	+10	100	1·6667
Not absent	130	140	−10	100	0·7143
Department 3:					
Absent	82	90	−8	64	0·7111
Not absent	218	210	+8	64	0·3048
				Therefore χ^2 =	3·5873

(*c*) *Degrees of freedom in a contingency table.* To find the degrees of freedom in a contingency table simply eliminate the last non-total row and the last non-total column and the number of remaining cells are the number of degrees of freedom in the table. In our example only two cells remain, and so there are only two degrees of freedom.

Department	Absent	Not Absent
1	✓	X
2	✓	X
3	X	X

(*d*) ***Interpreting the χ^2 result.*** If we want to use the 95 per cent level of confidence we look up in the χ^2 tables the cut-off χ^2 value for two degrees of freedom at 95 per cent level of confidence (*see* Appendix V). This value will be seen to be 5.991. This means our absenteeism χ^2 figure is below the cut-off point—i.e. the differences between departments are not significant and could easily have arisen by chance.

(However, as in the case of the non-rejection of the null hypothesis—*see* XXI, 4—this does not prove there is no difference between the departments, *only that the figures do not prove a difference.*)

7. Minimum expected cell values. One minor point needs making at this stage. If any of the *expected* cell values (observed values are not relevant here) are less than 5, then the χ^2 test tends to become inaccurate. In such a case the difficulty is surmounted by merging the "below 5" cells with adjoining ones so that the combined expected values in all the resulting cells are 5 or more.

8. "Expected value" formula. Sometimes the student is not too sure as to how the expected value for a given cell can be deduced. In the event of such uncertainty he can, if he wishes, apply the following formula:

$$\text{Expected cell value} = \frac{\text{row total} \times \text{column total}}{\text{grand total}}$$ i.e. multiply the total of the row the cell is in by the total of the column the cell is in and divide by the grand total of the whole table.

EXAMPLE

The "Department 2/Not absent" cell in our contingency table in *6*(*a*) will have an expected value of $\frac{200 \times 420}{600} = 140$.

9. Simple contingency tables. The student should note that on occasions a test involving but a single column can be made. For example, assume that there were four factories employing equal numbers of employees and that these four factories reported 107, 135, 116 and 94 accidents respectively during a given month. Do these differences indicate a real difference between the accident rates in the four factories?

Clearly, if there is *no* difference between the accident rates then we would expect an equal number of accidents in each factory. This means that since there is a total of 107 + 135 + 116 + 94 = 452 accidents in the four factories we would expect $\frac{452}{4} = 113$ accidents per factory. Our χ^2, then, would equal

$$\sum \frac{(\text{Actual accidents} - 113)^2}{113}$$

$$= \frac{(107 - 113)^2 + (135 - 113)^2 + (116 - 113)^2 + (94 - 113)^2}{113}$$

$$= 7{\cdot}8761.$$

The next step involves finding the degrees of freedom. In this sort of situation the degrees of freedom are equal to the number of different values less 1. So here there are 4 — 1 = 3 degrees of freedom. Looking up in our table, the value of χ^2 at the 95 per cent level of confidence where there are 3 degrees of freedom, we have a cut-off value of 7.815. Comparing this with our actual χ^2 value of 7.8761 shows that the actual observations just differ significantly from the expected values and so it is very probable that there is a real difference between the factories as regards their accident rates.

CURVE FITTING

NOTE: This section requires a knowledge of the probability distributions discussed in Chapter XVIII.

In this section we are concerned with seeing whether or not a mathematical curve can be made to fit a given set of observations. First we will see how to fit such a curve and then how to test if the fit is good or not.

10. Fitting a curve. Fitting a curve to a set of observations essentially involves plotting the observations on a graph and then trying to devise some mathematical equation which, when graphed, results in a curve that passes through, or close to, the plotted observations. Thus, to take a trivial case, imagine that we had the following observations:

Units of electricity:	1000	2000	3000	4000	5000
Electricity invoice:	£15	£25	£35	£45	£55

If one plots these observations on a graph it is quickly seen that the equation "Electricity invoice = £5 + (1p × number of units)" will give a curve that passes exactly through all the observed points. We can, then, fit this curve to the observations.

11. The curve equation. In most cases finding a suitable equation is not as easy as in the above instance. However, one may well have a hunch as to what the equation may be. For example, in a small maternity hospital the number of births each day would fluctuate but a little reflection would suggest that there was a good possibility that such births followed a Poisson distribution (Poisson distributions were discussed in Chapter XVIII). So we may well find that it is possible to fit a Poisson distribution to the number of daily births.

Suspecting the kind of equation that will fit the observations is, of course, only half the story—we also need to know what values must be slotted into the equation in order that we can get the curve we want. Thus, the Poisson distribution equation tells us that the probability of x occurrences $= e^{-a} \dfrac{a^x}{x!}$ but we still need to know what value a has to take.

Now in curve fitting any unknowns such as a are found from the observed data itself. In our example, since a represents the average number of births per day we must obviously have to find the average number of *observed* births. Slotting this average into the Poisson distribution equation will then enable us to find a theoretical set of figures that hopefully will fit the observed data.

EXAMPLE

Assume that our observed data were as follows:

Number of births during day:	0	1	2	3	4	5	6
Number of days:	171	290	273	152	72	32	10

In order to find a, the average number of births per day, from these observations, we must first find the total days (which is a simple addition of the second line) and then the total number of births (which involves multiplying each number of births during the day by the associated number of days and adding):

								Total
Number of days:	171	290	273	152	72	32	10	1000
Number of births:								
During day:	0	1	2	3	4	5	6	
During whole period:	0	290	546	456	288	160	60	1800

Since 1800 births occurred during 1000 days, the daily average number of births was $\frac{1800}{1000} = 1.8$. We can, then, substitute 1.8 for *a* in the Poisson distribution equation. This gives:

$$P(x) = e^{-1.8} \frac{1.8^x}{x!},$$

where P(*x*) is the probability of *x* births in a day. So to find the theoretical *number* of births in a day when there were 1000 days we must multiply P(*x*) by 1000.

The full calculation, therefore, reveals the following birth pattern:

No. of births during day	*No. of days*		
	Poisson distribution		*Actual for comparison*
0	$e^{-1.8} \times \frac{1.8^0}{0!} \times 1000 =$	165	171
1	$e^{-1.8} \times \frac{1.8^1}{1!} \times 1000 =$	297	290
2	$e^{-1.8} \times \frac{1.8^2}{2!} \times 1000 =$	267	273
3	$e^{-1.8} \times \frac{1.8^3}{3!} \times 1000 =$	160	152
4	$e^{-1.8} \times \frac{1.8^4}{4!} \times 1000 =$	72	72
5	$e^{-1.8} \times \frac{1.8^5}{5!} \times 1000 =$	26	32
6	$e^{-1.8} \times \frac{1.8^6}{6!} \times 1000 =$	8	10
More than 6	Balance =	5	0
		1000	1000

The visual comparison of expected with actual observations is shown in Fig. 40. As can be seen, the fit appears to be very good.

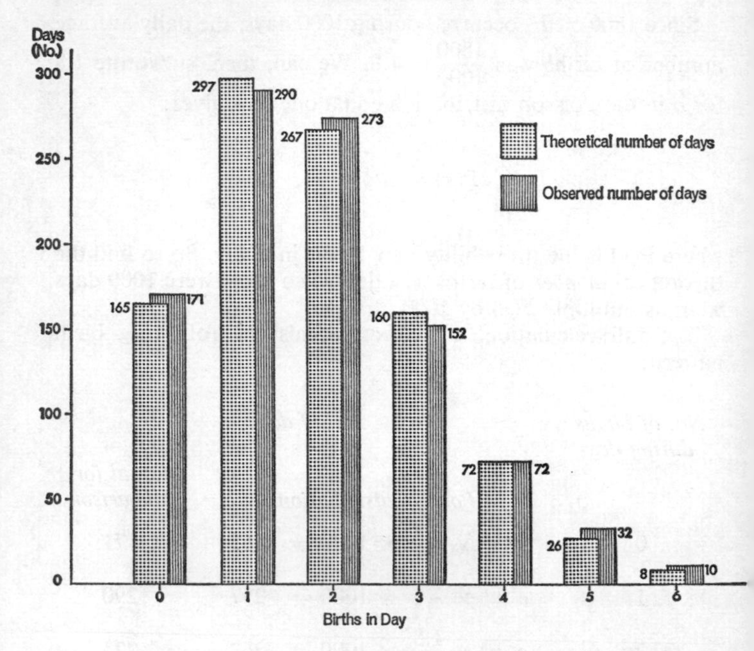

FIG. 40 *Fitting a Poisson distribution to an observed birth pattern.*

12. Testing the fit. After fitting a curve it is quite simple to test if the fit is good or not (i.e. to test if the mathematical equation does describe the underlying relationship and if the differences from the actual observations can be ascribed to chance)—it is merely a matter of applying a χ^2 test. The procedure is as follows:

(*a*) Find χ^2 from the χ^2 formula.
(*b*) Determine the degrees of freedom involved.
(*c*) Test the significance of the χ^2 found in (*a*).

EXAMPLE

Test the fit of the curve in the previous paragraph.

(a) Finding χ^2

Number of births during day	*Actual observations*	*Expected values*	$\frac{(\textit{Actual} - \textit{Expected value})^2}{\textit{Expected value}}$
0	171	165	0.2182
1	290	297	0.1650
2	273	267	0.1348
3	152	160	0.4000
4	72	72	0
5	32	26	1.3846
6	10	8	0.5000
More than 6	0	5	5.0000
			$\therefore \chi^2 = 7.8026$

(b) Determining the degrees of freedom. Finding the degrees of freedom in a curve-fitting exercise can be rather tricky. Essentially:

Degrees of freedom = (No. of classes used to find χ^2) − (No. of different figures taken from the observations in order to compute the expected values.)

Now the latter number in the above formula depends upon just how the theoretical curve has been computed. Clearly, to make the curve fit an actual set of observations, it is necessary to use in the equation one or more figures based on the observations and it is the number of such figures that must be deducted from the number of classes. In our example, for instance, we obtained the expected values by applying the formula:

$$\text{Expected value} = e^{-1.8} \times \frac{1.8^x}{x!} \times 1000,$$

where the 1.8 and the 1000 were obtained from the actual observations. Two different figures, then, were taken from the observations and since the number of classes were 8, the degrees of freedom = 8 − 2 = 6.

(c) *Testing the significance of the* χ^2. Tables show us that the value of χ^2 at the 95 per cent level of confidence where there are 6 degrees of freedom is 12.592. Since our value of 7.8026 is considerably less than this, there is no significant difference between the theoretical values and the actual observations. It is possible to conclude, therefore, that the births in the hospital very probably follow a Poisson distribution.

PROGRESS TEST 23

(*Answers in Appendix VI*)

1. Quality ratings were made at random on a number of production orders in two similar factories, X and Y. In X, 26 ratings were poor, 35 medium and 19 good; while in Y, 24 were poor, 55 medium and 41 good. Is the quality of the work done in Y significantly different from that in X?

2. A student of a game, in which scores range between 0 and 6 points, has observed that when played between teams organised into leagues the relative distribution of points scored per game during a season can be expressed by the formula:

No. of games recording a score of x points $= 4a - (a - x)^2$, where $a =$ average points per game over the whole season.

During last season the actual distribution of points in the top league was as follows:

Points	0	1	2	3	4	5	6
Games	12	54	66	90	80	64	34

Fit the theoretical curve to this distribution and test for goodness of fit.

CHAPTER XXIV

Statistical Quality Control

INTRODUCTION

1. Statistical quality control. *Quality control* is concerned with ensuring that the quality of goods or services remains within predetermined quality standards. *Statistical quality control* is a technique that employs statistical methods as an aid to controlling quality.

2. The function of statistical quality control. On first reading, the term "statistical quality control" seems to imply a technique concerned with indicating when quality ceases to be of the required standard. This is not strictly so, and it is important to appreciate this. The function of statistical quality control is to *signal when there is a quality change*. Basically the idea is this: if a producer sets up his process so that output of the required quality emerges and then uses statistical quality control to signal quality changes, it means that as long as no such signals are received the producer knows the quality of the output is of the required standard.

3. The need for statistical quality control. The question might well be asked, why bother about statistical quality control since a quality change can surely be detected simply by examining the output?

Unfortunately the matter is not so simple as this. Examination of output may well reveal a slight quality difference from the initial quality level but this may or may not be significant. There are two quite distinct reasons for such a difference, namely:

(*a*) Random effects causing a temporary quality change;
(*b*) Something in the process permanently altering so that the quality changes permanently.

Now random effects are almost always present. They include such things as small fluctuations in the process temperature, trivial instrument mis-readings by the operator, minor speed changes, slight tool movement, etc. Clearly there is no point in trying to adjust the process to allow for random effects—by definition they are transitory and affect the quality equally either way. On the other hand, if something in the process has permanently altered care must be taken that sub-standard production does not begin to emerge, and this may well mean stopping and adjusting the process.

The problem is, however, how can an inspector determine whether a slight quality change is due to *a random effect or a permanent process change*? Statistical quality control was developed to solve this problem.

4. Statistical basis of quality control. The two statistical quality control techniques we shall examine in this chapter are both simply practical applications of testing a hypothesis. Before reading further, then, the student should revise Chapter XXI.

A revision of this chapter will remind the student that the procedure differs slightly depending on whether one is testing a hypothesis relating to a mean or a proportion. This division of procedure has its counterpart in quality control so that there are "variable control charts" and "number defective control charts." These will now be examined in turn.

VARIABLE CONTROL CHARTS

5. Basic theory. We know from the theory of testing a hypothesis in relation to a mean (*see* XXI, **2**) that if a random sample is taken from a population then 19 times out of 20 the sample mean will be within two standard errors of the population mean. Envisage, then, a process which has been running for some time filling containers with an average 50 kg of material—though due to random effects most contain slightly more or less than 50 kg. It is reasonable in these circumstances to assert that *at this moment* the mean weight put into the containers is 50 kg and if we take a sample of containers being currently filled and find the sample mean weight of filling we would expect that 19 times out of 20 the sample mean would be within two standard errors of 50 kg. If the sample mean does lie within these limits, all well and good, but if it does not then either we have the exceptional 1 in 20 sample or *the process has permanently changed* so that it is no longer filling containers with 50 kg of material on average.

6. Application of basic theory. The application of this theory should be clear. If we keep taking samples and finding that their means lie within two standard errors of the original mean then it is probable all is well. However, if a sample mean exceeds these limits then it may indicate there has been a permanent change in the process giving rise to a quality change (the doubt being resolved if need be by taking a further sample at once and seeing if this confirms the suspected change or not).

7. Control limits. In statistical quality control the standard error limits discussed above are called *control limits.* Control limits are therefore simply levels of confidence. In practice 95 per cent control limits (called "warning" limits) and 99.8 per cent control limits (called "action" limits) are often used—a sample mean falling outside the action limits signalling a virtually certain quality change (since otherwise only once in 500 times could such a mean exceed these limits) and hence calling for action to re-adjust the process.

8. "In control." When the sample means lie within the control limits the process is said to be *in control.*

9. Specified values and control values. One thing often misunderstood by students is the relationship between the values specified by the designer and the control values. These are, in fact, quite different and the differences can perhaps best be explained by an example.

Assume a designer specifies that a metal rod must be cut 9 metres long, plus or minus 10 mm. We have, then, a *specified mean* of 9 m and *tolerances* of 8 m 990 mm and 9 m 10 mm.

Now note first that the random effects associated with our cutting machine must not cause variations in rod length exceeding a range of 20 mm, for if they did then no matter how carefully we set up the machine at the beginning, inevitably some of our rods will be defective. In such a situation new equipment is called for as statistical quality control is powerless here. However, let us assume the random effects only cause variations having a range of 6 mm. Now clearly this means we can start our machine cutting lengths of anything between 8 m 996 mm and 9 m 4 mm for at the worst (i.e. assuming our machine starts at the extreme end of the variation range) our rods will still be just within the tolerances, and it would, therefore, be a waste of time making further adjustments in order to obtain rods cut exactly 9 m long.

Finally, let us assume the machine starts by cutting rods on average 8 m 998 mm long with a standard deviation of 1 mm. (Note that since nearly all lengths will lie within three standard deviations from the mean, no defects will result from this setting.) If we now take samples of 100 then our standard error will be $1 \text{ mm}/\sqrt{100} = 0.1$ mm and 2 standard errors will be 0.2 mm. The 95 per cent control limit will, therefore, be 8 m 998 m $\pm$ 0.2 mm, i.e. 8 m 997.8 mm and 8 m 998.2 mm.

To sum up, then:

(*a*) The *specified mean* is 9 m.
(*b*) The *tolerances* are 8 m 990 mm and 9 m 10 mm.
(*c*) The *process mean* is 8 m 998 mm.
(*d*) The *control limits* (95 per cent) are 8 m 997.8 mm and 8 m 998.2 mm.

A little consideration of these figures will enable the student to appreciate the earlier remark that statistical quality control does not signal non-standard quality, only quality change, for here a change of the process mean to only 8 m 997.5 mm would be signalled virtually at once though all production at this new mean figure would still, of course, be well within the specified tolerances.

10. Practical difficulties. Although the above theory is sound as it stands there are a number of practical difficulties in its direct application. These are:

(*a*) Large samples take too long to measure. Often samples of only 4 or 5 items can be taken. This means using *t* distribution theory (*see* Chapter XXII).

(*b*) Computing the standard deviation of 4 or 5 items (to obtain the standard error) is too involved a procedure to be used on the shop-floor. Fortunately, there is a link between the standard deviation and the range (symbolised in quality control as *w*) for since three standard deviations either side the mean covers virtually the whole range of a distribution, then range $\simeq 6\sigma$, i.e. $\sigma \simeq$ Range/6 (though, again, small samples require some modification of this formula).

(*c*) Finally, as we saw in XIII, **2**, though the range is very easy to compute, it can be distorted by extreme values. This problem, however, can be overcome by taking some half dozen samples, *averaging their ranges*, and then using this average range ($\bar{w}$) in the formula in (*b*).

11. Control limit tables. The student must now accept that any required control limits can be computed from the following formula:

$$\textit{Control limits} = \textit{Process mean} \pm \bar{w} \times \mathrm{A}$$

where A is obtained from *control limit tables* which make allowances for all the considerations discussed in **10** above. These tables are very similar to *t* and χ^2 tables—in this instance, though, the left-hand column gives sample sizes while the other columns, each concerned with different confidence levels, give the appropriate values of A.

12. Charting the control data. Presentation of the control data is made on a chart which has time on the horizontal axis and the relevant range of variable values on most of the vertical axis (*see* Fig. 41). The initial process mean (found by computing the overall mean of the items in the initial group of samples taken to find $\bar{w}$) and the control limits are inserted as horizontal lines. Subsequent sample means are plotted against the time the samples are taken.

13. Summary—construction of a variable control chart. The procedure for constructing a variable control chart can now be summarised as follows (*see* **16** for illustration):

(*a*) Decide on a sample size (n).

(*b*) Take an initial half dozen or so samples of size n. From these find:

(*i*) the average range ($\bar{w}$)

(*ii*) the process mean ($\bar{x}$).

(*c*) Select appropriate confidence limits (e.g. 95 per cent and 99.8 per cent) and find associated control limits from the formula:

$$\textit{Control limits} = \bar{x} \pm \bar{w} \times \mathrm{A},$$

where A is taken from control limit tables.

(*d*) Construct the chart and insert horizontal lines for the process mean and the control limits.

(*e*) Take subsequent samples of size n at regular intervals, find the sample mean of each sample and plot against time the sample was taken.

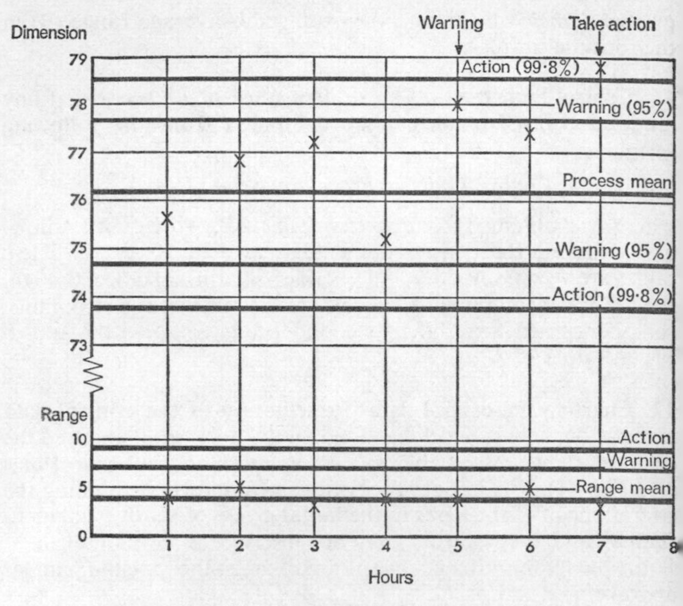

FIG. 41 *Variable control chart.*

For supporting figures *see* **16.** As long as the sample means fall within the control limits the process remains in control. (Note the inclusion in the lower part of the graph of a range control chart.)

14. Interpreting the chart. If the means of the samples taken subsequently to preparing the chart fall within the warning control lines then the process is in control. If a sample mean falls between the warning and action lines, then it means a possible change of quality has occurred.

NOTE: However, on average one sample mean in every 20 *will inevitably fall outside the warning lines* even if the process mean is unaltered, so once in a while we must expect this anyway.

If, though, a sample mean falls outside the action lines then the process is almost certainly out of control—i.e. a permanent change in quality has occurred.

15. Range charts. So far we have argued that as long as the means of samples taken from a process lie between the appropriate control limits then no significant quality change can have occurred. This, however, is not strictly accurate, for it is just possible that the dispersion of the process is increasing symmetrically—i.e. *the sample range is increasing but equally at each end so the mean remains unaltered.* As a result, although the mean remained the same, units could start being produced that had measurements outside the specified tolerances.

To check that this situation is not developing the lower part of the variable control chart is often adapted so that *sample ranges* can be recorded on the chart as well as the sample means. Similar theory to that discussed above exists to ensure that ranges are controlled along with the means.

16. Illustration of a variable control chart. To illustrate a variable control chart assume a process is started and six initial samples of 5 units each are taken from the output. Subsequently at 1 hour intervals further samples of 5 units are taken. Given the sample data opposite and that control limit tables show A to be 0.377 and 0.594 at the 95 per cent and 99.8 per cent levels of confidence respectively, then the resulting variable control chart is that shown in Fig. 41.

NUMBER DEFECTIVE CHARTS

17. Basic theory. Number defective charts are concerned with situations where an inspector simply accepts or rejects as defective an inspected unit. Here quality control is concerned with *proportions* of good and defective work and will therefore employ statistical theory relating to testing a hypothesis about proportions (*see* XXI, 9). We will therefore assert that the process proportion of defectives is x per cent, take a sample and find the sample proportion. If this proportion is within 2 standard errors of x per cent, well and good, but if it is not then either we have the exceptional 1 in 20 sample or the process proportion is not x per cent.

18. Application of the theory. The application of the theory closely follows that given in the previous section. First the actual proportion of defectives being produced is found and a check made to see this proportion meets a predetermined standard. Then control limits are computed. If subsequent sample proportions lie within the control limits, the process is in control —if not, the process may well be out of control.

	Initial samples						Subsequent hourly samples						
	A	B	C	D	E	F	1st	2nd	3rd	4th	5th	6th	7th
All measurements in cm	75	75	75	73	75	78	75	79	78	74	78	74	79
	78	74	75	76	79	73	73	74	78	76	76	79	80
	76	76	78	77	76	76	77	76	75	73	80	79	79
	78	79	77	76	78	75	76	76	77	76	79	76	77
	77	76	76	75	76	78	77	79	78	77	77	79	79
Total	384	380	381	377	384	380	378	384	386	376	390	387	394
$\bar{x}$	2286/30 = 76.2						75.6	76.8	77.2	75.2	78.0	77.4	78.8
Range	3	5	3	4	4	5	4	5	3	4	4	5	3
$\bar{w}$	24/6 = 4												

∴ Process mean = 76.2 cm
95% control limits = 76.2 cm ± 4 cm × 0.377* = 76.2 cm. ± 1.508 cm. = 74.692 cm. and 77.708 cm
99.8% control limits = 76.2 cm ± 4 cm × 0.594* = 76.2 cm ± 2.376 cm = 73.824 cm and 78.576 cm
Range control limits (theory not discussed):
95% limits = 4 cm × 1.81 = 7.24 cm
99.8% limits = 4 cm × 2.34 = 9.36 cm
* From control limit tables.

19. Why allow defects at all? At this point the student may well wonder why a business allows any defects at all. The reason is that it may be very expensive to eliminate totally a small proportion of defectives, particularly if the units are low-cost (e.g. imagine the expense involved in ensuring every screw had a proper slot in its head). Consequently it pays manufacturers to offer, and customers to accept, output having a specified maximum percentage of defectives at much reduced prices.

20. Practical considerations. To make the above theory practical it is necessary to take the following factors into consideration:

(*a*) Since small samples result in uselessly large standard errors, to obtain practical control limits it is necessary to take much larger samples than those taken for a variable control chart. However, inspection is much quicker in this situation as gauges can often enable an inspector to identify rapidly the defectives in a large batch of units.

(*b*) To avoid calculating the sample proportion every time a point is plotted on the control chart, the chart is constructed in terms of "number defective" (e.g. if 2 per cent is the process proportion and samples of 300 are to be taken, the chart would show the process "proportion" as 6 defectives in number). Note that the "number defective" can always be computed as follows:

$$\begin{aligned} \textit{Number defective} &= \textit{Sample size} \times \textit{proportion defective} \\ &= np \text{ (where } p = \text{proportion defective)} \end{aligned}$$

21. Control limits. To compute the control limits first note that if p = proportion defective, then the standard error of proportion defective $= \sqrt{\frac{pq}{n}}$ (*see* XX, 12). Therefore the standard error in terms of number defective

$$= \sqrt{\frac{pq}{n}} \times n = \sqrt{npq}.$$

The control limits, therefore, can now be found by application of the normal curve values to these expressions, i.e.

Control limits $= np \pm \sqrt{npq} \times$ *appropriate normal curve value*

(e.g. the 95% control limits would be $np \pm \sqrt{npq} \times 2$)

NOTE: The lower control limit is rarely used in practice since we are primarily interested in adverse quality changes only.

22. Finding the initial process number defectives. To start the control chart with the initial number of defectives being produced by the process, simply take a larger than normal sample, find the proportion defective and then multiply by the normal sample size to obtain the initial number defective.

23. Summary—construction of number defective control chart. The procedure for constructing a number defective control chart can be summarised as follows:

(*a*) Select a sample size consistent with the level of control sensitivity required (n).

(*b*) Take a larger than normal sample, find the proportion of defectives (p) and compute initial process defectives (np).

(*c*) Compute upper control limit from the formula:

Upper control limit $= np + \sqrt{npq} \times$
appropriate normal curve value for required confidence level

(*d*) Construct the number defective control chart having time on the horizontal axis and "number defective" on the vertical axis.

(*e*) Take subsequent samples of size n at regular intervals and plot the number defective in each sample against the time the sample was taken.

24. Interpreting the number defective control chart. The number defective control chart is, of course, interpreted in virtually the same way as a variable control chart (*see* **14**).

PROGRESS TEST 24

(*Answer in Appendix VI*)

1. A sample of 1000 items was taken from the initial output of a process. Thirty-five items were found to be defective. As this was well below the specified maximum defective level the process was allowed to run on. Subsequently samples of 400 were taken hourly, the first eight samples having 18, 11, 15, 10, 17, 14, 21 and 24 defectives respectively.

Construct a number defective chart, having 95 per cent and 99 per cent control limits, and plot the sample results for the first 8 hours. Make such comments as you feel necessary.

PART SEVEN

INDEX NUMBERS AND TIME SERIES

CHAPTER XXV

Index Numbers

In Part Three we emphasised the valuable function of statistics in reducing a mass of data, with the aid of measures, to a form easier to grasp. We now return to this aspect of statistics.

THE THEORY OF INDEX NUMBERS

1. The concept of an index number. If we wish to compare several series of figures it is more than likely that their complexity will render direct comparison meaningless. If, for instance, we had information on every form of production during this year and last year (*e.g.* tonnes of steel produced, litres of paint blended, TV sets manufactured, cars assembled and so on), the sheer mass of data would make it impossible to "see" in which year production was higher. Instead of such an embarrassing excess of figures, what we need is a *single* figure which in itself shows how much one year differs from another. A convenient way of doing it is to take a fairly typical year's figures as a base, and express the figures for other years as a percentage of this. Hence if the figure for 1975 were 100 and that for 1976 105, we should know that production (or whatever) was 5 per cent greater.

Such a single figure summarising a comparison between two sets of figures is called an *index number*.

2. Complications. In the first part of Chapter XII it was pointed out that arriving at a single figure to stand for a host of others results in many different opinions as to which figure serves best.

With index numbers this feature is again present, with the result that there emerges a confusing variety of types of index number and methods of calculation. Moreover the very act of compromise inevitably results in doubt as to the worth of the index number when it has been obtained. In this chapter only the minimum number of types will be examined and their worth left unquestioned.

3. Index number symbols. All the different methods by which index numbers can be calculated can be expressed concisely and unambiguously as formulae, and the student is therefore advised to refer to I, **9**, where index number symbols are explained. They should all be quite clear, except possibly the phrase *base year*, which can be defined as *the year against which all other years are compared.*

4. Base 100. In many ordinary day-to-day comparisons 100 is used as a base, percentages being the most obvious example. In consequence, people have become used to such comparisons and statisticians take advantage of this fact by basing index numbers on 100.

For example, if the production of TV sets was 38 261 last year and 43 911 this year we could call last year's production 100 and this years' (by simple proportion) 115. In this way the comparison between the two years' production is made much clearer.

A few indexes use other bases, such as 10 or 1000 but they are exceptional and there are usually good reasons for such a departure from normal practice (see *Index Numbers* by W. R. Crowe, Macdonald & Evans, 1965).

5. One-item index numbers. Where only one item is involved in comparisons between different periods, the calculation of index numbers is very simple. One year is chosen as base, and the values for other years are stated in proportion to the value of the base year, i.e.:

$$\text{Quantity index} = \frac{q_1}{q_0} \times 100\text{; or}$$

$$\text{Price index} = \frac{p_1}{p_0} \times 100$$

EXAMPLE

With 1972 as base year, compute quantity and price indexes for the years 1970 to 1976:

Year	*Price* (£)	*TV sets sold*	*Price index*	*Quantity index*
1970	45	12 912	$\frac{45}{50} \times 100 = 90$	$\frac{12\,912}{21\,200} = 61$

1971	48	18 671	$\frac{48}{50} \times 100 = 96$	$\frac{18\,671}{21\,200} = 88$
1972	50	21 200	etc. **100**	etc. **100**
1973	53	28 633	106	135
1974	53	35 028	106	165
1975	55	40 650	110	192
1976	60	44 531	120	210

NOTE: A one-item price index is also called a *price relative*; *see* **15**.

WEIGHTED AGGREGATIVE INDEXES

6. Need for multi-item indexes. Unfortunately, index numbers are often wanted in circumstances where there is more than just one item. To take a common instance, we often need an index number that compares the cost of living in one year with that in another. Clearly, more than one item is involved in the cost of living!

For demonstration purposes, let us assume that only three items enter a cost-of-living index—bread, cheese and ale—and that prices in the two years to be considered in an historical survey were:

TABLE XI. DATA FOR THREE-ITEM COST OF LIVING INDEX

Item	*1965*	*1970*
Bread	4½p loaf	5p loaf
Cheese	50p kg	100p kg
Ale	90p keg	75p keg

7. Difficulties involved in multi-item indexes. Our aim now is to determine a single figure which will compare the cost of living in 1970 with that of 1965. Examination of the figures above reveals at least three difficulties:

(*a*) Two prices have gone up and one down. As there can only be a single index number *it must be a compromise* between these two opposing price movements.

(*b*) The prices are given *for different units*. It is not feasible, therefore, to add together all the prices for a single year.

(*c*) There is no indication as to *how important* each item is in the cost of living. Obviously, bread should be considered more important than ale.

Difficulty (*a*) is a feature of index numbers that must always be borne in mind. Index numbers *are* compromises.

8. Weighting. Difficulties (*b*) and (*c*), on the other hand, can be overcome by *weighting*, i.e. multiplying the price by a number (the *weight*) that will adjust the item's value in proportion to its importance. For example, if cheese is given a weight of 10, its original price of 50p becomes a weighted price of 500p (£5.00).

When such a weight is selected, both the importance of the item and the unit in which the price is expressed are taken into consideration. Consequently, weighted figures are directly comparable.

Assume that weights of 100, 10 and 1 are given respectively to bread, cheese and ale. The weighted figures will therefore be:

TABLE XII. WEIGHTED COST OF LIVING INDEX

Item	*Weight*	1965		1970	
		Price	*Price × weight*	*Price*	*Price × weight*
Bread	100	$4\frac{1}{2}$p	450	5p	500
Cheese	10	50p	500	100p	1000
Ale	1	90p	90	75p	75
Total			1040		1575

It is now possible to compute a single index number simply by calling the total of the 1965 weighted price column 100, and finding the total of the 1970 column as a proportion, i.e.:

$$\text{Index number for 1970 (1965} = 100) = \frac{1575}{1040} \times 100 = \underline{\underline{151}}.$$

9. Summary of procedure.

(*a*) List the items and prices.
(*b*) Select weights.
(*c*) Multiply the prices by selected weights (*weight*).
(*d*) Add the products (*aggregate*).
(*e*) Compare the total for the base year with the total for the other year by using percentages.

From this summary it will be clear why an index computed by this method is called a *weighted aggregative index*.

10. Formula for a weighted aggregative index. The formula for calculating such an index is:

$$Index = \frac{\Sigma (p_1 \times w)}{\Sigma (p_0 \times w)} \times 100$$

LASPEYRE AND PAASCHE INDEXES

11. Quantity weighted indexes. In the previous section the actual selection of weights for a weighted aggregative index was not discussed. In practice it can be a difficult problem. One solution is to use the *actual quantities consumed* as weights. Obviously, the more bread that is consumed the more important bread is as a cost of living item. Similarly, the low consumption of, say, caviare would reflect the insignificance of such an item. An index number so computed is known as a *quantity weighted index*. Such an index differs from the one discussed above only in the use of actual quantities for weights.

12. Base year or current year quantities? Using actual quantities is all very well, but the question immediately arises as to *which* quantities? Those consumed in the base year, or those consumed in the year for which the index is required?

The answer is that either can be used—although, of course, different index numbers are obtained as a result. It so happens that each method is named after its original inventor. The one which uses base year quantities is called a *Laspeyre* index, and the one that uses the current year quantities a *Paasche* index.

13. Definitions and formula.

(*a*) *Laspeyre price index:* a base year quantity weighted index. The formula is:

$$Index = \frac{\Sigma p_1q_0}{\Sigma p_0q_0} \times 100$$

A Laspeyre price index indicates how much the cost of buying base-year quantities at current-year prices is, compared with base-year costs.

(*b*) *Paasche price index:* a current-year quantity weighted index. The formula is:

$$Index = \frac{\Sigma p_1q_1}{\Sigma p_0q_1} \times 100$$

A Paasche price index indicates how much current-year costs are related to the cost of buying current-year quantities at base-year prices.

EXAMPLE

Compute (1) *Laspeyre and* (2) *Paasche price indexes of the following data* (1965 = *base year*):

Item	1965		1970	
	Price (p_0)	*Quantity* (q_0)	*Price* (p_1)	*Quantity* (q_1)
Bread	4½p loaf	80 000 loaves	5p loaf	100 000 loaves
Cheese	50p kg	10 000 kg	100p kg	15 000 kg
Ale	90p keg	1 000 kegs	75p keg	3 000 kegs

(1) *Laspeyre price index:*

Item	p_0	p_1	q_0	$p_0 \times q_0$ (£)	$p_1 \times q_0$ (£)
Bread	4½p	5p	80 000	3 600	4 000
Cheese	50p	100p	10 000	5 000	1 000
Ale	90p	75p	1 000	900	750
				9 500	14 750
				$\Sigma p_0 q_0$	$\Sigma p_1 q_0$

Using the Laspeyre formula, Index $= \frac{14\,750}{9500} \times 100 = 155$

(2) *Paasche price index*

Item	p_0	p_1	q_1	$p_0 \times q_1$ (£)	$p_1 \times q_1$ (£)
Bread	4½p	5p	100 000	4 500	5 000
Cheese	50p	100p	15 000	7 500	15 000
Ale	90p	75p	3 000	2 700	2 250
				14 700	22 250
				$\Sigma p_0 q_1$	$\Sigma p_1 q_1$

Using the Paasche formula, Index $= \frac{22\,250}{14\,700} \times 100 = 151$

14. Laspeyre and Paasche indexes contrasted. The next query a student may raise is, what difference does it make which index is chosen? As regards the final figure, there will probably be very little difference unless there has been a substantial change in the purchasing pattern. There are, however, two important practical points involving the computation and use of these indexes:

(*a*) Paasche numbers require actual quantities to be ascertained for *each* year of the series. This can be a big requirement. In contrast, a Laspeyre index requires quantities for the base year only.

(*b*) With Paasche numbers, the denominator of the formula, Σp_0q_1, needs recomputing *every year*, as q_1 changes yearly. In the case of Laspeyre numbers, however, the denominator, $\Sigma\, p_0q_0$, always remains the same. Moreover, a consequence of this is that different years in a Laspeyre index can be directly compared with each other, whereas in a Paasche series the changing denominator means that different years can be compared *only* with the base year and not with each other.

For these reasons Laspeyre indexes are much more common than Paasche indexes.

OTHER INDEXES

15. Weighted average of price relatives index. A *price relative* is simply the price of an item in one year relative to another year—again expressed with 100 as base. Symbolically, it is

$$\frac{p_1}{p_0} \times 100$$

Thus, bread in our earlier example being 4½p in 1965 and 5p in 1970, has a price relative of

$$\frac{5\text{p}}{4\frac{1}{2}\text{p}} \times 100 = \underline{\underline{111}}$$

Since each price relative is, in effect, a little one-item index number (*see* 5), a composite index number can be obtained by *averaging* all the price relatives of items in a series. Again, weighting is necessary to allow for item importance.

EXAMPLE

Find the weighted average of price relatives index figures in Table XI with weights of 10, 7 and 3 respectively:

Item	*1965 price*	*1970 price*	*Price relative*	*Weight*	*Price relative × weight*
Bread	4½p	5p	111	10	1110
Cheese	50p	100p	200	7	1400
Ale	90p	75p	83	3	249
				20	2759

$$\therefore \text{ Index} = \frac{2759}{20} = \underline{\underline{138}}$$

NOTE: Remember that in a weighted average you divide by the sum of the weights (*see* XII, **12**).

The formula, therefore, is:

$$\textit{Weighted average of price relatives index} = \frac{\sum\left(\frac{p_1}{p_0} \times 100 \times w\right)}{\Sigma w}$$

It should be noted, incidentally, that the spread of weights in this index is much smaller than that of the weighted aggregative index. On the face of it this suggests that the drop in price of ale would have a greater influence. Yet the index is higher than before! The reason is that a price relative is quite different from a price: small prices, for instance, can have large relatives if they are unstable. For this reason, indexes using price relatives must not be quantity-weighted (though they can be value-weighted).

16. Chain index numbers. A *chain index* is simply an ordinary index in which each period in the series uses the *previous period as base*. For instance, a simple example of a one-item chain index, using some of the data given in **5**, is given in Table XIII.

Such an index shows whether the *rate* of change is rising (rising numbers), falling (falling numbers) or constant (constant numbers) as well as the *extent* of the change from year to year. In the example just given, it can be seen that although there is a steady increase in the sales of television sets the increase each year, in relation to the total sales of the previous year, is on the whole falling.

TABLE XIII. ONE-ITEM CHAIN INDEX

Sales of television sets

Year	*TV sets sold*	*Chain index*
1970	12 912	
1971	18 671	$\frac{18\,671}{12\,912} \times 100 = 145$
1972	21 200	$\frac{21\,200}{18\,671} \times 100 = 114$
1973	28 633	$\frac{28\,633}{21\,200} \times 100 = 135$
1974	35 028	$\frac{35\,028}{28\,633} \times 100 = 122$
1975	40 650	$\frac{40\,650}{35\,028} \times 100 = 116$
1976	44 531	$\frac{44\,531}{40\,650} \times 100 = 110$

In a multi-item index such as one measuring the cost of living, a chain index is useful inasmuch as new items can be introduced. For instance, if it is wished to introduce continental holidays into such an index this year, then data for this year and last year only is required. Had a normal, non-chain index been used based on (say) 1948, there would probably have been no appropriate data available for that year and so it would be impossible to introduce continental holidays into the index.

17. Quantity indexes. Apart from the indexes of television set sales in **5** and Table XIII every index examined in this chapter has been a *price index*, i.e. a measure of price changes. There are other kinds of indexes. An obvious one is an index of quantity.

A *quantity index* is one that measures changes in quantities. There are virtually as many different methods of computing a quantity index as there are a price index—in fact, formulae for quantity indexes can be derived from those for price indexes by simply interchanging the p and q symbols. Thus, a Laspeyre price index with a formula of

$$\frac{\Sigma\, p_1 q_0}{\Sigma\, p_0 q_0} \times 100$$

would become a Laspeyre *quantity* index with a formula of

$$\frac{\Sigma\, q_1 p_0}{\Sigma\, q_0 p_0} \times 100$$

Weighting is particularly necessary when constructing a multi-item quantity index as it is otherwise impossible to add together kilogrammes, litres, pairs, etc.

18. Value indexes. Another group of indexes relate to *value*, value being, of course, $p \times q$. Thus

$$\frac{\Sigma\, p_1 q_1}{\Sigma\, p_0 q_0} \times 100$$

is a *value index* since it compares values in the base year with values in a subsequent year.

CHANGING THE BASE

19. When base-changing necessary. It sometimes happens that the user of an index wishes to change the base year. This often happens when two different series are to be compared, since it is unlikely that both will have the same base year, and so direct comparison between them would be difficult.

EXAMPLE

Assume that an index of new television licences for the years 1970–76, with 1960 as base year, ran as follows:

Year:	1970	1971	1972	1973	1974	1975	1976
Index:	210	230	250	300	360	410	500

Direct comparison with the index of television-set sales computed in **5** is hardly possible. To obtain such a direct comparison it is necessary to change the base year of one of the series so that both have the same base.

20. Procedure for changing the base. The procedure is as follows:

(*a*) look up the index number relating to the new base year, then

(*b*) divide this number into each index number in the series and multiply by 100.

This will give a new series of index numbers with the new year as its base.

NOTE: Changing the base of a weighted index gives a series slightly different from that which would be obtained if the index had been computed entirely afresh with the new year as base. But for practical purposes the difference is rarely significant.

In the above example, it means that to change the index of new television licences from a base year of 1960 to a base of 1972, all the figures will have to be divided by 250 (the 1972 index number). To illustrate this, all the data relating to this example may be tabulated:

Year	*TV sets sold* (1972 = 100) (*see* 5)	*TV licenses* (1960 = 100)	*TV licences* (1972 = 100)
1970	61	210	$\frac{210}{250} \times 100 = 84$
1971	88	230	$\frac{230}{250} \times 100 = 92$
1972	**100**	250	$\frac{250}{250} \times 100 = \mathbf{100}$
1973	135	300	$\frac{300}{250} \times 100 = 120$
1974	165	360	$\frac{360}{250} \times 100 = 144$
1975	192	410	$\frac{410}{250} \times 100 = 164$
1976	210	500	$\frac{500}{250} \times 100 = 200$

Comparison of the two series is now possible and it indicates that while the rate of increase in the sales of television sets was initially greater than the rate of increase in licences, the trend was reversed in the later part of the series. If the sales index relates to one company's sales, the figures indicate that it has begun to lose its share of the market.

INDEX CONSTRUCTION

21. Factors involved. When constructing an index number, four factors need to be considered:

(*a*) The purpose of the index.
(*b*) The selection of the items.
(*c*) The choice of weights.
(*d*) The choice of a base year.

22. Purpose of the index. The purpose of an index must be very carefully decided, for decisions relating to the other three factors will depend on the purpose. Moreover, the *interpretation* of the index will also depend on the purpose.

For example, an index constructed to measure change in building costs must not be used for revaluing machinery—nor even the commercial value of a building, since such an index would not take into account changes in the values of land on which such buildings were situated.

23. Selecting the items. This can be the most difficult problem of all. Take the construction of a cost-of-living index. Obviously, bread should be included, but what about table wines? Heating costs ought to be included, but how about television viewing costs? If home rentals are selected, should holiday rentals be selected too?

In the case of an index measuring employment, are part-time workers to be included? What about self-employed workers? In an export index, what should we do about imports which are immediately re-exported? Or imports returned for some reason to the overseas supplier? Whose share prices do we use for a share price index?

Moreover, the problem may arise as to *which* figures to take. Is a cost-of-living index to be based on prices in London or Manchester? Or in Little-Comely-on-the-Ouse?

The answers to such problems lie in defining the purpose of the index carefully and then selecting the items that will best achieve that purpose—although it must be realised there will always be differences of opinion.

Other problems in this category are the selection of items:

(*a*) *which are unambiguous:* an index of mortality from a given disease would be seriously distorted if improved diagnosis is attributing more and more deaths to the disease that previously had been attributed to other causes; and

(*b*) *whose values are ascertainable:* the construction of an index relating to undetected murders would run into obvious difficulties.

24. Choice of weights. The problem here is to find weights which will result in each item being given its appropriate importance. Actual quantities may often be good weights but they are not invariably appropriate. If it were decided to take the quantities used by a "typical household" as weights, there would be some difficulty in determining a "typical household." After all, spinsters on the old age pension are very much concerned with cost-of-living figures and they would hardly be impressed with a heavy weighting for, say, private car travel.

But there is one factor which makes the problem of choosing weights easier. This is that a difference of opinion as regards weights does not, oddly enough, affect the index as much as one would suppose. For instance, a completely revised weight of 200 for bread in Table XII (i.e. a doubling of the original weight—which was already over *nine* times as big as the other weights combined) results in an index of 139 as against a previous number of 151—a mere change of 8 per cent. Smaller revisions would result in smaller differences and this indicates that hair-splitting as regards weights is rarely worthwhile.

25. Choice of a base year. Generally speaking, the year chosen as base should be (*a*) a reasonably normal year, and (*b*) not too distant.

Sometimes a year which is significant within the series may be chosen; for example, the year a Commonwealth country attained its independence might be an appropriate base year for an index of that country's production, or the year of nationalisation might make a logical base year for U.K. coal output.

Choosing a freak year is a favourite trick of those who use statistics to mislead. A dishonest capitalist could choose a record year for profits as base and so "prove" subsequent profits to be pitifully low. A dishonest trade unionist could similarly choose a year of exceptionally full employment to "prove" that current unemployment is intolerably high.

CHOOSING AN INDEX

26. Which index should be used? To this question there can only be one answer: *it depends on the circumstances.* The Laspeyre index is a good all-round one but it cannot be used if weighting figures (normally quantity) are unobtainable. In that case, the weighted average of price relatives, with weights determined on some other basis, may be appropriate. The purpose of the index is highly relevant. A chain index is obviously called for when the purpose is to indicate to what extent figures have changed in relation to the previous year. A chain index is valuable, too, where new items may need to be added and old items removed.

Finally, it must be emphasised once more that an index number is only an attempt to summarise a whole mass of data in one figure. Such a figure must inevitably be subject to many limitations and it is the responsibility of the user to balance all factors and judge:

(*a*) which type of index is appropriate, and

(*b*) the *real* significance of any single index number in the series.

PROGRESS TEST 25

(*Answers in Appendix VI*)

1. From the data below, and using 1973 as base where appropriate,

(*a*) Draw up:

(*i*) A Laspeyre price index.

(*ii*) A Paasche price index.

(*iii*) A weighted average of price relatives, using the weighting:

A, 5; B, 3; C, 2.

(*b*) Compute:*

(*i*) A quantity index for A alone.

(*ii*) A price index for B alone.

(*iii*) A quantity chain index for C alone.

(*iv*) A Laspeyre quantity index.

(*v*) (1) An index having the formula:

$$Index = \frac{\Sigma p_1 q_1}{\Sigma p_0 q_0}.$$

(2) What type of index is this?

	1973		1974		1975		1976	
Item	*Price* £	*Quantity*	*Price* £	*Quantity*	*Price* £	*Quantity*	*Price* £	*Quantity*
A	0.20	20	0.25	24	0.35	20	0.50	18
B	0.25	12	0.25	16	0.10	20	0.12½	16
C	1.00	3	2.00	2	2.00	3	2.00	4

CHAPTER XXVI

Time Series

INTRODUCTION

1. What is a time series? Many variables have values that change with time, e.g. population, exports, car registrations, company sales, employment and electricity demand. Figures relating to the changing values of a variable over a period of time are called a *time series*. For example, Table XIV is a time series showing company sales changing over time.

TABLE XIV. TIME SERIES

Sales of PQP Co. Ltd, 1973–76 (tonnes)

Year	*Quarter 1*	*Quarter 2*	*Quarter 3*	*Quarter 4*	*Total*
1973	672	636	680	704	2692
1974	744	700	756	784	2984
1975	828	800	840	880	3348
1976	936	860	944	972	3712

2. Factors influencing a time series. If a graph is drawn of a time series (e.g. Fig. 42 relating to Table XIV), the following features can often be seen:

(*a*) *Seasonal variation.*—A regular up-and-down pattern that repeats *annually* and is due to the effect of seasons on the variable.

(*b*) *Cyclical variation.*—A regular up-and-down pattern that repeats over a span of years. In the main it reflects the boom/depression economic cycle.

NOTE: As the boom/depression cyclical variation is nowadays much more dependent on government policy than economic rhythms, there is rarely any underlying *time-based* pattern to be analysed. This form of variation is, then, ignored in the rest of this chapter.

(*c*) *The trend.*—An overall tendency for the curve to rise (or fall).

(*d*) *Random* (*or residual*) *variation.*—Odd movements of the curve which fit into no pattern at all.

Each of these factors affects the curve and, because they all do so simultaneously, it is difficult to distinguish clearly the influences of any single one. For instance, in Table XIV it appears that seasonal influences make the fourth quarter the busiest in the year. In actual fact, the first quarter is the busiest (as will be shown later), but the combined effect of seasonal influence and trend disguise this. In order to determine the influence of each factor, it is necessary to isolate each in turn.

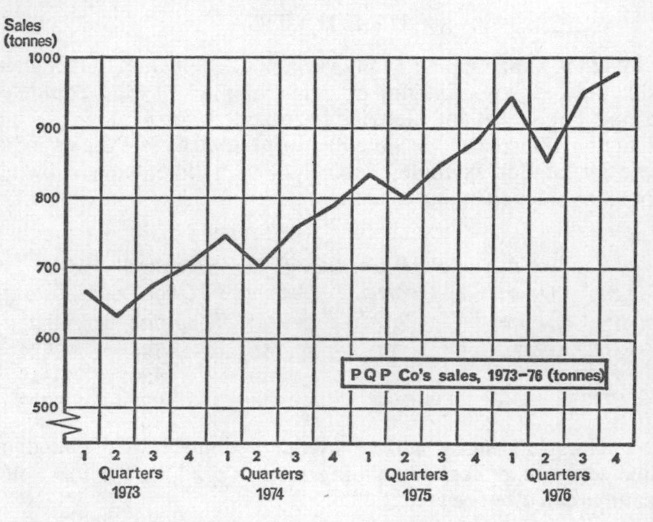

FIG. 42 *Time series: Table XIV*

3. Object of the exercise. There are two very good reasons why it is worth while isolating each factor, for if their individual influences are known we are able to:

(*a*) *Predict future values of the variable.* Such knowledge is of great value. For example, if the total demand for electricity 10 years from now can be estimated, the facilities that will be needed in 10 years and which have a 10-year constructional period can be started at once, not in (say) 6 years' time, when it will be too late.

(*b*) *Control events.* In exercising control it is often very important to know at the earliest possible moment should a new element enter the situation. The interaction of existing factors tends to hide the appearance of a new element until it has already had unforeseen effects. Analysis of the series helps to reveal "intruders" at an early stage. If, when the actual figures are received, they differ from the predicted ones by an amount greater than could be explained by random variation, there is a strong probability that a new influence has entered the series, altering either the trend or the seasonal pattern. In **8** an example is given where sales appear to be maintaining a continuous rise to the end of the year but in reality stopped rising 6 months earlier.

SEASONAL VARIATION

4. The method of moving averages. There are several ways of analysing a time series to isolate the seasonal variation, but probably the most generally satisfactory one is the *method of moving averages*, which involves computing moving averages (*see* VI). Each average, of course, eliminates seasonal influence and is located at the centre of the period it relates to. The figure for the season it is located against is then found as a percentage of the moving average, and this measures the variation caused by seasonal factors.

Unfortunately, if there is an even number of seasons, a moving average will inevitably be centred between two seasons. To overcome this difficulty the moving averages on either side of a particular season are averaged to give a *centred average* centred on that season. This will probably be better understood by studying the example used to illustrate the procedure given in **5**.

Examination of the resulting seasonal variation detailed in Table XV shows that there is a slight inconsistency. Quarter 2, for example, has figures of 94, 97 and 94 respectively. This is on account of random variations. To eliminate such variations, the seasonal variations are averaged:

Year	*Quarter 1*	*Quarter 2*	*Quarter 3*	*Quarter 4*	
1973	—	—	100	101	
1974	103	94	100	101	
1975	103	97	99	101	
1976	105	94	—	—	
Total	311	285	299	303	*Total*
Average	$103\frac{2}{3}$	95	$99\frac{2}{3}$	101	$= 399\frac{1}{3}$

TABLE XV. ANALYSIS OF SEASONAL VARIATION

Using data from Table XIV

Year	*Quarter*	*Actual sales*	*M.A.T.*	*Moving average*	*Centred average*	*Seasonal variation*
1973	1	672				§
	2	636				§
			2692*	673		
	3	680			682†	100‡
			2764	691		
	4	704			699	101
			2828	707		
1974	1	744			716½	103
			2904	726		
	2	700			736	94
			2984	746		
	3	756			756½	100
			3068	767		
	4	784			779½	101
			3168	792		
1975	1	828			802½	103
			3252	813		
	2	800			825	97
			3348	837		
	3	840			850½	99
			3456	864		
	4	880			871½	101
			3516	879		
1976	1	936			892	105
			3620	905		
	2	860			916½	94
			3712	928		
	3	944				§
	4	972				§

* This M.A.T. figure was found by adding the sales of the four quarters of 1973. Since this is only a step on the way to finding a moving average it is located at the mid-point of this period, i.e. between quarters 2 and 3.

† Centred average: mean of two moving averages on either side of it.

‡ Seasonal variation:

$$\frac{\textit{Actual figure}}{\textit{Centred average}} \times 100$$

(Be careful not to work with this formula upside down.)

§ No figures possible for these quarters.

It will be noted that the four quarters do not add up to 400. Averaging the seasonal variations of the quarters frequently leads to this result, but since the total of the four quarterly variations must equal 400 (*see* 6) an adjustment is needed to bring this about. This adjustment is often made by adding to each seasonal variation a fraction equal to:

$$\frac{\textit{Required total} - \textit{Actual total}}{\textit{Number of seasons}}$$

i.e. in this case,

$$\frac{400 - 399\frac{1}{3}}{4} = \frac{1}{6}$$

If this formula results in a minus, subtract the adjustment from the variations.

NOTE: Strictly speaking, the larger variations should carry a proportionately larger share of the adjustment, though usually the amount involved is too small to warrant such a refinement.

When a fraction as small as this is obtained, an arbitrary adjustment is more reasonable. In this example it is probably best to round the two seasonal variations with fractions to the next complete number. This would result in the final seasonal variations being:

Quarter 1	*Quarter 2*	*Quarter 3*	*Quarter 4*	*Total*
104	95	100	101	= 400

5. Summary of procedure: method of moving averages.

(*a*) This procedure is illustrated in Table XV. Lists the series vertically.

(*b*) Compute the moving totals and insert these at the mid-points of the relevant periods.

(*c*) Compute the moving averages.

(*d*) If there is an even number of seasons, average the adjacent moving averages to give centred averages centred on each season.

(*e*) Compute (*Actual figure* ÷ *centred average*) × 100. This gives the individual seasonal variations.

(*f*) Find the mean of the individual seasonal variations for each season.

(*g*) Adjust these means so that the sum of *all* seasonal variations is 100 × *Number of seasons*. The figures arrived at after this adjustment are the final seasonal variations.

It should be noted that this procedure can be applied to any kind of season. If months are used, the only effects are that the calculations are larger and that the adjustment in step (*g*) would entail making the sum of the seasonal variations 1200. Note too, that seasons need not be seasons of a year. Morning, afternoon and night are "seasons" in relation to electricity demands, for instance; so are certain times of day for passenger transport services.

6. Theory underlying the method of moving averages. From the computations above, quarter 1 emerges as the busiest season, with a seasonal variation of 104. Examination of the original figures in Table XIV suggests that quarter 4 should be the busiest. The contradiction arises because in this series there is a strong upward trend: if there were no seasonal variations at all, the last quarter would be distinctly higher than the first because, coming later in time, it benefits from the trend.

In the series we have examined, the trend is so steep that although quarter 4 was not as busy seasonally as quarter 1, sales in the last quarter were always higher than in the first quarter of the same year.

It is in order to eliminate the distortion of the seasonal figures by the trend that the method of moving averages is used. The approach is to find out what the figure for each season would be if there were no seasonal variation, and then relate the actual figure to this as a percentage. In VI we saw that moving averages eliminated seasonal variations and therefore such averages are used.

The adjustment in step (*g*) is needed because if some seasons are above average (where the average is 100) some must be below and, of course,

$$\frac{\Sigma\ (\textit{Seasonal variations})}{\textit{Number of seasons}}$$

must equal 100. It follows that the sum of the seasonal variations must be 100 × *Number of seasons.*

DESEASONALISING A TIME SERIES

7. Why deseasonalise? When examining figures subject to seasonal variation, one of the problems is to know whether (say) a relatively high figure in a busy season is due wholly to seasonal factors, or whether some other factors are involved. If the seasonal variations are known, they can be used to remove the seasonal influences from the figures. The resulting figures are then said to be *deseasonalised* (or *seasonally adjusted*). The influence of factors other than seasonal variations can then be seen.

8. Computation of deseasonalised figures. The formula is:

$$\textit{Deseasonalised figure} = \frac{\textit{Actual figure}}{\textit{Seasonal variation}} \times 100$$

EXAMPLE

Assume that when the 1977 sales figures relating to Table XIV were received they were:

Quarter 1: 1020 tonnes. Quarter 3: 1010 tonnes.
Quarter 2: 960 tonnes. Quarter 4: 1020 tonnes.

What are the deseasonalised figures and what conclusions can be drawn from them?

The deseasonalised figures are found as follows:

Quarter	*Actual figure*	*Seasonal variation*	*Deseasonalised figure*
1	1020	104	$\frac{1020}{104} \times 100 = 980$
2	960	95	$\frac{960}{95} \times 100 = 1010$
3	1010	100	$\frac{1010}{100} \times 100 = 1010$
4	1020	101	$\frac{1020}{101} \times 100 = 1010$

The conclusion we can draw is that the upward trend has levelled out. From the original 1977 data it seemed as though sales were on a downward trend in quarter 2 and returned to an upward trend in quarters 3 and 4. The deseasonalised figures show that in actual fact the upward trend was maintained during quarter 2, and in quarters 3 and 4 the trend levelled out.

THE TREND AND THE METHOD OF SEMI-AVERAGES

9. Finding the trend. Now that we have seen how to isolate the seasonal variation, we can examine the next factor, the *trend*.

Since moving averages indicate trend, it might be thought that the trend of any series would be given by the fifth column of Table XV. Unfortunately, moving averages also include random variations along with the trend, so they do not result in the straight trend lines needed for prediction and control.

There are two principal ways of finding the trend. They are:

(*a*) The method of semi-averages.
(*b*) The method of least squares.

Remember, incidentally, that when graphing any trend figures it is necessary always to plot totals at the *mid-point* of the period to which they refer.

10. Features of the method of semi-averages. This is by far the easier of the two methods of finding the trend, but it is rather crude and is apt to be inaccurate if there are any extreme values in the series. However, providing these limitations are borne in mind, it can be usefully employed in appropriate circumstances.

11. Procedure with the method of semi-averages.

(*a*) Compute the annual totals (or totals of complete cycles, should the seasons be other than of the normal yearly kind).

(*b*) Divide the series into two halves, each containing a complete number of years. If the overall series contains an odd number of years, omit the middle year.

(*c*) Compute the mean value of each half.

(*d*) Plot these two mean values on a graph at the mid-points of their respective periods and join the points. This gives the *trend line.*

EXAMPLE

The trend of the series in Table XIV, using the method of semi-averages, is found as follows:

SALES TREND OF DATA IN TABLE XIV

	Year	*Total sales for year (tonnes)*	*Semi-average*	*Mid-point of period*
	1973	2692	2838	End 1973
	1974	2984		
Dividing line				
	1975	3348	3530	End 1975
	1976	3712		

12. Using the trend line for prediction. If the trend line is projected to the right, an estimate of future yearly totals can be read

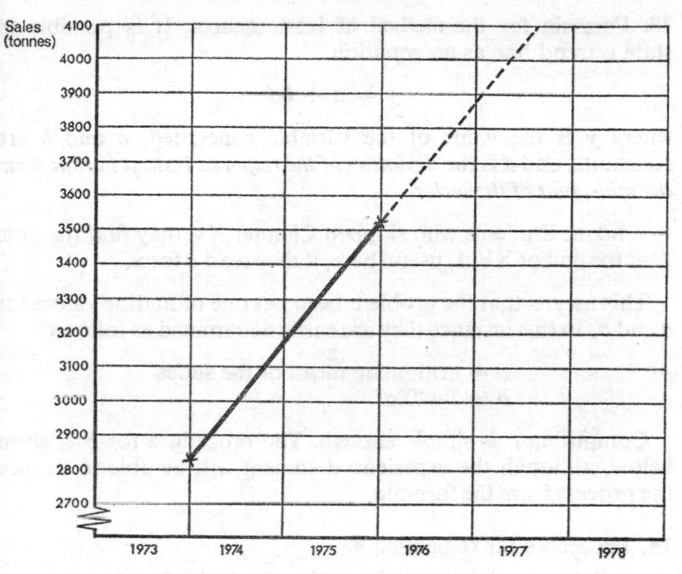

FIG. 43 *Method of semi-averages.*

from it. It must be borne in mind, though, that on this particular type of graph, annual totals must be read at the mid-point of the year concerned. Thus, in Fig. 43, for example, the estimate for sales in 1977 is 4050 tonnes.

THE TREND AND THE METHOD OF LEAST SQUARES

Students who studied Chapter XV will remember the method of least squares. A moment's reflection will show that a trend line is, in fact, an ordinary regression line of the variable concerned on time (since we wish to predict variable value from time and not vice versa). This means that the student who wishes to use the method of least squares to find a trend line may compute the regression line in the ordinary way and completely ignore the procedure outlined in this section. However, it may pay him to read what follows, for a simpler method of computing the regression line is possible in a time series, owing to the fact that time values increase in equal increments.

13. Formula for the method of least squares. It is possible to state a trend line as an equation,

$$y = a + bd$$

where y is the value of the variable concerned, a and b are constants, and d is the *deviation of the required instant in time from the mid-point of the series.*

NOTE: Students who skipped Chapter XV may find the note at the end of XV, 1, useful here, if they read d for x.

This means that the problem becomes one of finding values for a and b. In this instance they are easily determined as follows:

$$a = \text{arithmetic mean of the series}$$
$$b = \Sigma yd/\Sigma d^2$$

Computing a is simple enough. The procedure for b is given below, although the experienced student will be able to deduce the process from the formula.

14. Procedure for computing *b*.

(*a*) Set down the annual figures in a vertical column.
(*b*) Find the mid-point of the series.
(*c*) Against each year, insert the deviation of the mid-point of that year from the mid-point of the series. (With an odd number of years in the series these deviations will be whole years, but with an even number they will involve half years.)
(*d*) Multiply the variable value for each year by its deviation and add the products to give Σyd.
(*e*) Square the deviations and add the products to give Σd^2.
(*f*) Apply formula given in **13**.

EXAMPLE

Find the trend line of the data in Table XIV.

Year	*y (sales)*	*d*	*yd*	d^2
1973	2692	−1.5	−4038	2.25
1974	2984	−0.5	−1492	0.25
Mid-point				
1975	3348	+0.5	+1674	0.25
1976	3712	+1.5	+5568	2.25
	$\Sigma y =$ 12 736		$\Sigma yd =$ +1712	$\Sigma d^2 =$ 5.00

Now $a = \text{mean of the series} = \frac{\Sigma y}{n} = \frac{12\,736}{4} = \underline{\underline{3184}}$

and $b = \Sigma yd/\Sigma d^2 = \frac{+1712}{5} = \underline{\underline{342.4}}$

$\therefore$ Trend: $\underline{\underline{y = 3184 + 342.4d}}$

NOTE: The mid-point of the series is the *end* of 1974. On the other hand each year's figures are, of course, *centred on the middle of the year.*

15. Use of the computed trend line. Having calculated the trend line, prediction is possible in one of two ways:

(*a*) The trend line can be graphed by taking any two years and:

(*i*) computing their deviations d_1 and d_2;

(*ii*) inserting these values of d_1 and d_2 in the trend formula and finding y_1 and y_2;

(*iii*) plotting d_1, y_1 and d_2, y_2 on the graph;

(*iv*) joining the two points to obtain a trend line. The trend line can then be extended and the estimated values in future years read off directly.

(*b*) Much more simply, take the year for which a prediction is required and:

(*i*) find its deviation d from the mid-point of the original series;

(*ii*) insert this value in the trend formula and compute y.

EXAMPLE

Estimate the sales for 1977 from the data in Table XIV.

(*a*) 1977 has a deviation d of 2.5 years from the mid-point of the series in Table XIV.

(*b*) $\therefore\ y = a + bd = 3184 + 342.4 \times 2.5 = 4040$ tonnes.

NOTE: Here the method of least squares gives to all intents and purposes the same answer as the method of semi-averages, because the series of figures is a simple series and the added sophistication of the least squares method is not really needed.

FORECASTING

16. Forecasting future values. Having isolated the seasonal variation and the trend in a time series, we are left only with the random variations. Deferring consideration of these for the moment, it can doubtless be seen that if the trend and seasonal variations are known, then it is possible to compute *future* values of the variable (allowing also, of course, for "cyclical" variation in the form of future government policy and world economic trends). All one needs to do, in fact, is to compute the trend value for any future period and adjust for the seasonal variation.

17. Subdividing the trend. Whilst the principle is very simple, its practical execution is complicated slightly by the need to subdivide the annual trend figure into figures for each seasonal period. If we call the trend figure for the *seasonal* period y', then the trend formula becomes:

$$y' = \frac{a}{n} + \frac{b}{n^2} \times d'$$

where n = number of seasons per annum, and d' = deviations measured in these periods.

The trend figure for each seasonal period may now be computed by this formula.

Special care has to be taken in using the formula to count d' correctly. Remember that figures are centred at the mid-points of their periods, i.e. seasonal figures are centred at the mid-point of the seasons. Thus, if the deviation of quarter 3, 1976 was required in our Table XIV example, it would be determined as follows:

Number of full quarters from mid-point of the series (end of quarter 4, 1974) to end of quarter 2, 1976 $= 6$
$+$ half a quarter for quarter 3, 1976 $= \frac{1}{2}$

$\therefore d' = 6\frac{1}{2}$

18. Applying the seasonal variation. The next step after computing the trend figures for each individual period is to apply the seasonal variation. This simply entails multiplying the trend figure by the seasonal variation. This gives the required forecasted figures.

19. Random variations. The figure forecasted in the previous paragraph will, of course, exclude the *random variations*. However, by definition, these are unforecastable. Nevertheless it is valuable to have some idea of how far a forecast is likely to be in error because of a random variation.

Finding this is not particularly difficult. All one needs to do is to use the forecasting formulae to "forecast" *past* values and then compare these "forecasts" with the actual values. The differences will indicate the extent to which the random variations are influencing the actual results. If these differences are very small then the forecast will probably be very accurate (providing no new factor makes its entry), and vice versa.

EXAMPLE

Find to what extent random variations affected the figures in Table XIV.

The annual trend was found to be:

$y = 3184 + 342.4\,d$ (where d = deviation in years).

Therefore the *quarterly* trend will be:

$$y' = \frac{3184}{4} + \frac{342.4}{4^2} \times d' = 796 + 21.4d'$$

where d' = deviation in quarters.

The computation of the random variation is, therefore, as follows:

Quarter		d'	Trend (from formula above)	Seasonal variation 4	Forecasted figures (Trend × Seasonal variation)	Actual figures	Random variation: Actual − Forecast	Random variation: %
1973	1	$-7\frac{1}{2}$	636	104	660	672	12	+2
	2	$-6\frac{1}{2}$	657	95	624	636	12	+2
	3	$-5\frac{1}{2}$	678	100	678	680	2	0
	4	$-4\frac{1}{2}$	700	101	707	704	−3	0
1974	1	$-3\frac{1}{2}$	721	104	749	744	−5	−1
	2	$-2\frac{1}{2}$	742	95	706	700	−6	−1
	3	$-1\frac{1}{2}$	764	100	764	756	−8	−1
	4	$-\frac{1}{2}$	785	101	793	784	−9	−1
Mid point								
1975	1	$+\frac{1}{2}$	807	104	839	828	−11	−1
	2	$+1\frac{1}{2}$	828	95	787	800	13	+2
	3	$+2\frac{1}{2}$	850	100	850	840	−10	−1
	4	$+3\frac{1}{2}$	871	101	880	880	0	0
1976	1	$+4\frac{1}{2}$	892	104	928	936	8	+1
	2	$+5\frac{1}{2}$	914	95	868	860	−8	−1
	3	$+6\frac{1}{2}$	935	100	935	944	9	+1
	4	$+7\frac{1}{2}$	956	101	966	972	6	+1

These figures indicate that the random variation is very small, and the actual figures will probably be within 1 per cent or 2 per cent of the predicted figures—assuming that no new factor enters the series.

20. A forecast example. We are now in a position to prepare a forecast, and as an example let us assume we wish to forecast the 1977 sales of PQP Ltd.

Quarters 1, 2, 3 and 4 of 1977 have *d*'s of 8½, 9½, 10½ and 11½ respectively from the end of 1974, our trend line reference point. They also have seasonal variations of 104, 95, 100 and 101. Our forecast, then, will be prepared as follows:

Forecast for 1977 *Sales, PQP Ltd.* (*tonnes*)

Quarter	d'	*Trend* $796 + 21{\cdot}4d'$	*Seasonal variation*	*Forecast*
1	8½	978	104	1017
2	9½	999	95	949
3	10½	1021	100	1021
4	11½	1042	101	1052

If random variations keep within the previous limits then these forecasts should prove correct to within $\pm$ 2 per cent.

Forecasts for other periods could, of course, be made in a similar manner.

INTERPOLATION AND EXTRAPOLATION

21. Interpolation and extrapolation. A common technique in statistics is to plot a series of points on a graph and then draw a line of best fit across the graph. It was used in connection with scattergraphs, regression lines and can now be applied to time series trends. The line of best fit is used to predict values, and it is in this context that interpolation and extrapolation arise.

(*a*) *Interpolation* consists in reading a value on that part of the line which lies *between* the two extreme points plotted (i.e. a value on the continuous line in Fig. 44).

(*b*) *Extrapolation* means reading a value on the part of the line that lies *outside* the two extreme points plotted (i.e. a value on the dotted line in Fig. 44).

The distinction between the two is necessary, for although interpolation is permissible, it is considered dangerous to extrapolate. In the case of interpolation, the actual points on the graph give a sound indication of the possible error that could arise in reading a value from the line. But, where the line lies *outside* the plotted points, there is no guide at all to the degree of error. Although the plotted points may suggest the line has a steady slope, it may well be that some new, unsuspected factor comes into play at the higher or lower levels which, unknown to us, alters the slope in those regions.

For example, a scattergraph of heights and ages where ages are below 16 years would indicate continuous growth. If the height of an 80 year-old man was estimated from the line of best fit of such a scattergraph, the estimate would be something like 7.5 m!

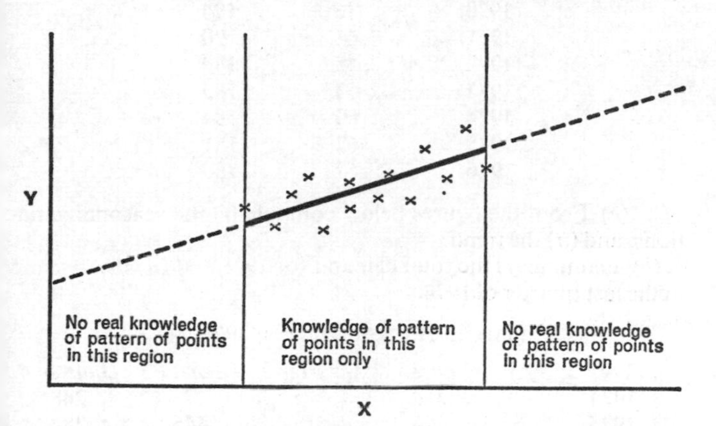

FIG. 44 *Interpolation and extrapolation.*

In this example, we know from experience that growth stops at about 16 years of age and that extrapolation would result in serious error. But in many cases we have no experience of the events lying beyond the two extremes and so therefore do not know if extrapolating is allowable or not. It is true that the estimate may be reasonably accurate if it is read from a part of the line lying beyond, *but still close to*, one of the extreme points; but the further one moves away from the extreme point the more one must treat the estimate with caution.

PROGRESS TEST 26

(*Answers in Appendix VI*)

1. The data below relate to the populations, in thousands, of two towns A and B, between 1968 and 1976.

(*a*) Compute the population trend line for each town.

(*b*) Estimate when the populations of the two towns will be equal, and state the estimated size of the populations at that time.

Year	*A*	*B*
1968	25	200
1969	31	196
1970	36	194
1971	44	190
1972	48	189
1973	53	185
1974	60	184
1975	62	181
1976	67	180

2. (*a*) From the figures below compute (*i*) the seasonal variations, and (*ii*) the trend.

(*b*) Estimate (*i*) the total demand for 1978 and (*ii*) the demand in the last quarter of 1978.

DEMAND FIGURES (TONNES) 1974–76

Year	*Quarter* 1	*Quarter* 2	*Quarter* 3	*Quarter* 4
1974	218	325	273	248
1975	444	585	445	385
1976	660	852	623	525

3. The following figures relate to units of service demanded of a service enterprise working 24 hours a day.*

(*a*) Find (*i*) the seasonal variation, (*ii*) the trend, and (*iii*) the random variation.

(*b*) Graph the actual figures and superimpose the trend line.

Day	*Morning*	*Afternoon*	*Night*
1	820	310	600
2	800	330	600
3	860	340	700
4	900	380	680

4. (*For students who have studied regression lines* (*Chapter XV*).) Prove that a trend line found by the method of least squares is the regression line of y on x, where x is time (*Hint:* If deviations are measured from the mid-point of a series and such deviations are in equal steps, note that the sum of these deviations must be zero.)

APPENDIX I

Bibliography

For further study, the student is referred to the following books:

Freund and Williams, *Modern business statistics*, Pitman, 1970.

For more detailed discussion on statistical data (*e.g.* economic, social and vital statistics, and surveys): A. R. Ilersic, *Statistics*, HFL Publishers Ltd, 1964.

For more emphasis on the mathematical aspects of statistics: Brookes and Dick, *Introduction to statistical method*, Heinemann, 1969.

For a popular book on the use of statistics—which also goes beyond the present work: M. J. Moroney, *Facts from figures*, Pelican, 1969.

For the next step: R. Goodman, *Teach yourself statistics*, English Universities Press.

For two light but informative books on the interpretation of statistics: D. Huff, *How to lie with statistics*, Gollancz, 1954, and *Fred Learns Basic Statistics*, Continua Productions, 1975.

For a slightly more expanded section on probability theory: W. M. Harper, *Operational Research*, Macdonald & Evans, 1975.

For fuller treatment of index numbers: W. R. Crowe, *Index numbers: theory and applications*, Macdonald & Evans, 1965.

And for further reading: G. L. Thirkettle, *Wheldon's business statistics and statistical method*, Macdonald & Evans, 1976.

APPENDIX II

Statistical Formulae

ESSENTIAL FORMULAE

Frequency distributions

Arithmetic mean ($\bar{x}$)

"Direct method" formula: $\bar{x} = \dfrac{\Sigma x}{n}$

"Short method" formula:

$$\bar{x} = \text{Mid-point of chosen class} + \left(\frac{\Sigma fd}{\Sigma f} \times \text{Class interval}\right)$$

This formula is for use with frequency distributions that have equal class intervals where d = deviation in whole classes from the chosen class. If the distribution has unequal class intervals, or is an ungrouped distribution, then $d = (x - \text{Mid-point of chosen class})$ and "$\times$ Class interval" is dropped from the formula; *see* Appendix III.

Range

Range = Highest value − Lowest value

Quartile deviation

$$\text{Quartile deviation} = \frac{Q_3 - Q_1}{2}$$

Standard deviation (σ)

"Direct method" formula: $\sigma = \sqrt{\dfrac{\Sigma(x - \bar{x})^2}{n}}$

"Short method" formula:

$$\sigma = \sqrt{\frac{\Sigma fd^2}{\Sigma f} - \left(\frac{\Sigma fd}{\Sigma f}\right)^2} \times \text{Class interval}$$

See note to short method formula for arithmetic mean, above.

$$\textit{Variance} = \sigma^2$$

Coefficient of variation

$$\text{Coefficient of variation} = \frac{\sigma}{\bar{x}} \times 100$$

Pearson coefficient of skewness (Sk)

$$Sk = \frac{3(\bar{x} - \text{Median})}{\sigma}$$

Correlation

Correlation (r)

$$r = \frac{\Sigma xy - n\bar{x}\bar{y}}{n\sigma_x\sigma_y}$$

where n = the number of pairs.

Rank correlation (r′)

$$r' = 1 - \frac{6\ \Sigma d^2}{n(n^2 - 1)}$$

where n = the number of pairs, and
d = the difference between rankings of the same item in each series.

Standard errors

Standard error of the mean ($\sigma_{\bar{x}}$)

$$\sigma_{\bar{x}} = \frac{\sigma_{sample}}{\sqrt{n}}$$

Standard error of a proportion (σ_p)

$$\sigma_p = \sqrt{\frac{pq}{n}}$$

where p = sample proportion, and
$q = 1 - p$.

Standard error of the difference between means ($\sigma_{(\bar{x}_1 - \bar{x}_2)}$)

$$\sigma_{(\bar{x}_1 - \bar{x}_2)} = \sqrt{\sigma^2_{\bar{x}_1} + \sigma^2_{\bar{x}_2}}$$

Standard error of the differences between proportions ($\sigma_{(p_1 - p_2)}$)

$$\sigma_{(p_1 - p_2)} = \sqrt{\sigma^2_{p_1} + \sigma^2_{p_2}}$$

χ^2 distribution

$$\chi^2 = \sum \frac{(\text{Actual} - \text{Expected Value})^2}{\text{Expected Value}}$$

Index numbers

Weighted aggregative price index

$$\text{Index} = \frac{\Sigma\,(p_1 \times w)}{\Sigma\,(p_0 \times w)} \times 100$$

Laspeyre price index

$$\text{Index} = \frac{\Sigma\,(p_1 \times q_0)}{\Sigma\,(p_0 \times q_0)} \times 100$$

Paasche price index

$$\text{Index} = \frac{\Sigma\,(p_1 \times q_1)}{\Sigma\,(p_0 \times q_1)} \times 100$$

USEFUL FORMULAE

Weighted average

$$\text{Weighted average} = \frac{\Sigma\,xw}{\Sigma w}$$

Geometric mean

$$\text{GM} = \sqrt[n]{x_1 \times x_2 \times x_3 \times \ldots\ldots x_n}$$

$$\left(\text{alternative: Log GM} = \frac{\Sigma \log x}{n}\right)$$

Harmonic mean

$$\text{HM} = \frac{n}{\Sigma\frac{1}{x}}$$

Regression lines

Line equation: $y = a + bx$

Regression line of y on x:

$$\left.\begin{aligned}\Sigma y &= an + b\Sigma x\\ \Sigma xy &= a\Sigma x + b\Sigma x^2\end{aligned}\right\}\text{Solve for } a \text{ and } b$$

where n = the number of pairs.

Regression line of x on y:

Interchange x and y in the above simultaneous equations. (*see* XV, 5).

Quality control

$$\text{Control limits (process mean)} = \bar{x} \pm \bar{w} \times A$$

where A is taken from control limit tables

Control limits (number defective) $= np \pm \sqrt{npq}\ \times$ Appropriate Normal Curve value for required confidence

Regression coefficient

The value of b in the $y = a + bx$ regression line equation.

Bayes' Theorem

$$\text{P(Prior event } E|\text{Subsequent event } S) = \frac{\text{Probability of } E \times \text{P}(S|E)}{\text{Probability of } S \text{ one way or another}}$$

Standard normal distribution

$$z = \frac{x - \bar{x}}{\sigma}$$

Price relative

$$\text{Price relative} = \frac{p_1}{p_0} \times 100$$

Weighted average of price relatives index

$$\text{Index} = \frac{\sum\left(\frac{p_1}{p_0} \times 100 \times w\right)}{\Sigma w}$$

Base changing

$$\text{New index number} = \frac{\text{Old index No.}}{\text{Old index No. of new base period}} \times 100$$

Deseasonalised figures

$$\text{Deseasonalised figure} = \frac{\text{Actual figure}}{\text{Seasonal variation}} \times 100$$

Trend line

$$y = a + bd$$

where y = the variable for which trend is required
d = the deviation in time from the mid-point of the time series

$$a = \bar{y}, \text{ and } b = \frac{\Sigma yd}{\Sigma d^2}$$

Subdivision of the trend line

$$\text{Sub-period value } y' = \frac{a}{n} + \frac{b}{n^2} d'$$

where n = the number of sub-periods per cycle, and
d' = the deviation from the mid-point of the series measured in sub-periods.

APPENDIX III

Computational Methods

MEANS, STANDARD DEVIATIONS AND MEAN DEVIATIONS IN GROUPED FREQUENCY DISTRIBUTIONS

When the "direct method" of computing a mean or standard deviation would result in involved calculations an alternative method (called a "short method") can be used. Here are detailed the steps taken in the three most commonly used short methods. (Note that the "mid-point of chosen class" is technically called the "assumed mean.")

Method 1: To find mean and standard deviation of an equal class interval distribution.

Application. To be used to find $\bar{x}$ and σ for frequency distributions having *equal class intervals.*

Steps.

1. Lay out table as follows:

Column no.:	1	2	3	4	5
Column title:	*Class*	f	d	fd	fd^2
How found:	Given	Given	See rule below*	Col. 2 × Col. 3 (Watch plus and minus signs)	Col. 3 × Col. 4 (Note minus signs disappear)
Total		Σf		Σfd	Σfd^2

* RULE: To find d, first choose any class you wish. The "d" for this chosen class is 0. Then, in the "d" column, number off all classes above and below the chosen class—those above being given minus signs and those below plus signs (*see* example).

2. *To find* $\bar{x}$. Apply following formula:

$$\bar{x} = \text{Mid-point of chosen class} + \left(\frac{\Sigma fd}{\Sigma f} \times \text{Class interval}\right)$$

(Note that if Σfd is negative, it will mean the bracketed amount will be *subtracted* from the mid-point of chosen class.)

3. *To find* σ. Apply following formula:

$$\sigma = \sqrt{\frac{\Sigma fd^2}{\Sigma f} - \left(\frac{\Sigma fd}{\Sigma f}\right)^2} \times \text{Class interval}$$

(NOTE: You multiply by the class interval *after* finding square root of rest of expression.)

GENERAL NOTE. Remember that $\bar{x}$ and σ will be in the same units as the original data—and these units should be stated in answer (e.g. if the original data was in inches, then $\bar{x}$ and σ will be in inches).

EXAMPLE

Find $\bar{x}$ and σ of following distribution (Table VIII):

Kilometres	*f*
400–under 420	12
420– „ 440	27
440– „ 460	34
460– „ 480	24
480– „ 500	15
500– „ 520	8

1. Layout is as follows (selecting "440–under 460" as the chosen class):

Class (km)	*f*	*d*	*fd*	*fd*²
400–under 420	12	−2	−24	48
420– „ 440	27	−1	−27	27
440– „ 460	34	0	0	0
460– „ 480	24	+1	+24	24
480– „ 500	15	+2	+30	60
500– „ 520	8	+3	+24	72
Σ	120		+27	231

2. $\bar{x}$ = mid-point of chosen class + $\left(\frac{\Sigma fd}{\Sigma f} \times \text{Class interval}\right)$

$$= 450 + \left(\frac{+27}{120} \times 20\right) = 450 + 4.5 = \underline{\underline{454.5 \text{ kilometres}}}$$

3. $$\sigma = \sqrt{\frac{\Sigma fd^2}{\Sigma f} - \left(\frac{\Sigma fd}{\Sigma f}\right)^2} \times \text{Class interval}$$

$$= \sqrt{\frac{231}{120} - \left(\frac{27}{120}\right)^2} \times 20 = \sqrt{1.92 - 0.05} \times 20$$

$$= \sqrt{1.87} \times 20 = 1.37 \times 20 = \underline{\underline{27 \text{ kilometres}}}$$

Method 2: To find mean and standard deviation of an unequal class interval distribution.

Application. To be used to find $\bar{x}$ and σ for grouped frequency distributions having *unequal class intervals.*

Steps.

1. As step 1 in method 1 *except* change "Rule" to:

RULE: To find *d*, first choose any class you wish. The "*d*" for this chosen class is 0. To find "*d*" for any other class apply following formula:

d = mid-point of other class − mid-point of chosen class (watch the sign of the answer).

2. *To find $\bar{x}$ and σ.* Apply method 1 formulae *except* "Class interval" must be dropped from formulae, i.e.:

$$\bar{x} = \text{mid-point of chosen class} + \frac{\Sigma fd}{\Sigma f}$$

$$\sigma = \sqrt{\frac{\Sigma fd^2}{\Sigma f} - \left(\frac{\Sigma fd}{\Sigma f}\right)^2}$$

EXAMPLE

Find $\bar{x}$ and σ of the following distribution:

£'s	*f*
25–under 50	30
50– „ 100	50
100– „ 200	35
200– „ 500	20
500– „ 1000	5

1. Layout is as follows (selecting "100–under 200" as the chosen class):

Class (£'s)	*f*	*Mid-point*	*d*	*fd*	*fd²*
25–under 50	30	37.5	−112.5	−3375	379 688
50– „ 100	50	75	−75	−3750	281 250
100– „ 200	35	150	0	0	0
200– „ 500	20	350	+200	+4000	800 000
500– „ 1000	5	750	+600	+3000	1 800 000
	140			−125	3 260 938

2. $\bar{x}$ = mid-point of chosen class $+ \dfrac{\Sigma fd}{\Sigma f}$

$$= 150 + \frac{-125}{140} = 150 - 0.89 = \underline{\underline{£149.11}}$$

$$\sigma = \sqrt{\frac{\Sigma fd^2}{\Sigma f} - \left(\frac{\Sigma fd}{\Sigma f}\right)^2} = \sqrt{\frac{3\,260\,938}{140} - \left(\frac{-125}{140}\right)^2}$$

$$= \sqrt{23\,291.6} = £152.6$$

Method 3: To find mean deviation of any grouped frequency distribution.

Application. To be used to find the mean deviation of any grouped frequency distribution.

Steps.

1. Add the following columns to the normal layout shown in methods 1 and 2 above:

....	*MP (mid-point) (if not already included)*	*(MP − $\bar{x}$)*	*f(MP − $\bar{x}$)*
How found:	Mid-point of class	Class mid-point less mean Note: IGNORE SIGN	Frequency × previous column
			$\Sigma f(MP - \bar{x})$

2. Find mean in normal manner and complete layout.
3. Apply formula: Mean deviation $= \frac{\Sigma f(MP - \bar{x})}{\Sigma f}$.

EXAMPLE

Find mean deviation of Table VIII detailed in method 1 above.

1.

Class (km)	*f*	*MP*	$(MP - \bar{x})$	$f(MP - \bar{x})$
400–under 420	12	410	44.5	534
420– „ 440	27	430	24.5	661.5
440– „ 460	34	450	4.5	153
460– „ 480	24	470	15.5	372
480– „ 500	15	490	35.5	532.5
500– „ 520	8	510	55.5	444
Σ	120			2697.0

2. From method 1 above $\bar{x} = 454.5$ kilometres.
3. Mean deviation $= \frac{\Sigma f(MP - \bar{x})}{\Sigma f}$

$$= \frac{2697}{120} = 22.48 \simeq 22\tfrac{1}{2} \text{ kilometres.}$$

STANDARD DEVIATIONS OF NON-GROUPED DISTRIBUTIONS

Method 4: To find standard deviation of a group of figures using an assumed mean.

Application. To be used when the mean of a group of figures is not a round number and laborious calculations would result if the direct method was used as explained in XIII, 6.

Steps.

1. Choose any number as the assumed mean.
2. Using the chosen number as $\bar{x}$, find $\Sigma(x - \bar{x})$ and $\Sigma(x - \bar{x})^2$.
3. Apply formula: $\sigma = \sqrt{\frac{\Sigma(x - \bar{x})^2}{n} - \left(\frac{\Sigma(x - \bar{x})}{n}\right)^2}$

(where n = number of figures in group).

EXAMPLE

Find σ of 5, 7, 8, 12 and 18 (i.e. the same figures as used in XIII, 6) using an assumed mean.

1. Choose (say) 8 as the assumed mean.
2. Layout is as follows:

x	$(x - \bar{x})$		$(x - \bar{x})^2$
5	$5 - 8$	-3	9
7	$7 - 8$	-1	1
8	$8 - 8$	0	0
12	$12 - 8$	$+4$	16
18	$18 - 8$	$+10$	100
	$\Sigma(x - \bar{x}) =$	$+10$	$\Sigma(x - \bar{x})^2 = 126$

3. $$\sigma = \sqrt{\frac{126}{5} - \left(\frac{+10}{5}\right)^2} = \sqrt{25.2 - 4}$$
$$= \sqrt{21.2} = 4.6$$

Method 5: To find standard deviation of a group of figures using a calculator.

Application. To be used when a calculator is available. The method is based on the fact that an alternative formula for finding a standard deviation of an ungrouped distribution is

$$\sigma = \sqrt{\frac{\Sigma x^2}{n} - \bar{x}^2}$$

Steps.

1. Add x's; Divide by n; Square; STORE.
2. Square x's and add progressively; Divide by n; Recall STORE and subtract stored value; Find square root; ANSWER $= \sigma$

EXAMPLE

Find standard deviation of 5, 7, 8, 12 and 18.

1. $5 + 7 + 8 + 12 + 18 = 50$; $\div 5 = 10$;
   Square $= 100$; STORE 100.
2. $5^2 + 7^2 + 8^2 + 12^2 + 18^2 = 606$; $\div 5 = 121.2$;
   $-$STORE $100 = 21.2$; $\sqrt{21.2} = 4.6 = \sigma$.

REGRESSION LINES AND *r*

Method 6: To find regression lines or r *using the "short" method.*

Application. To be used when the size of the figures involved would lead to cumbersome calculations. The method is based on the fact that any series of figures involved in a regression line or *r* calculation can have any constant added or subtracted and/or can be multiplied or divided by any constant without the computation procedure being affected.

Steps.

1. If desired, add/subtract any chosen constant to/from every figure in the 1st series.
2. If desired, multiply/divide every resulting figure in the 1st series by any other chosen constant.
3. Repeat steps 1 and 2 for the second series using same or different constants as desired.
4. Use adjusted figures in the requisite mathematical procedure.
5. Take the result given by 4 and in the case of:

(*a*) Regression line equation:

(*i*) Subject any given right-hand variable to the *same* adjustment as made to the series of that variable above.

(*ii*) Apply the step 4 equation.

(*iii*) Subject the answer, the estimated variable, to a *reversal* of the adjustment made to the series of that variable above.

(*b*) *r:* Make no compensating adjustment (*r* will be correct as it stands).

EXAMPLE

Find *y* on *x* regression line and *r* of following series:

x	y
1720	−333
2000	−441
2520	−621
2040	−405

1. Subtract 2000 from all *x* values.
2. Divide all resulting *x* values by 40.
3. Add 405 to all *y* values and then divide all resulting values by 18.

4.

x	y	x^2	xy	$(x-\bar{x})^2$	$(y-\bar{y})^2$
−7	4	49	−28	76.56	42.25
0	−2	0	0	3.06	0.25
13	−12	169	−156	126.56	90.25
1	0	1	0	0.56	6.25
7	−10	219	−184	206.74	139.00
÷ 4: 1.75	−2.5			51.69	34.75

(*a*) y on x regression line

(*i*) $\Sigma y = an + b\Sigma x$

(*ii*) $\Sigma xy = a\Sigma x + b\Sigma x^2$

(*i*) $-10 = 4a + 7b$

(*ii*) $-184 = 7a + 219b$

(*i*) $\times 7$: $-70 = 28a + 49b$

(*ii*) $\times 4$: $-736 = 28a + 876b$

Subtract: $666 = \quad - 827b$

$\therefore b = -0.8053$

So (*i*):

$-10 = 4a + (7 \times -0.8053)$

$\therefore a = -1.09$

$\therefore y = -1.09 - 0.8053x$

(*b*) r

$$r = \frac{xy - n\bar{x}\bar{y}}{n\sigma_x\sigma_y}$$

$$\therefore r = \frac{-184 - (4 \times 1.75 \times -2.5)}{4 \times \sqrt{51.69} \times \sqrt{34.75}} = -0.982$$

5. To find estimated y when x is, say, 2200:

(*i*) Adjustment of 2200:

$$\frac{2200 - 2000}{40} = 5$$

(*ii*) $\therefore y = -1.09 - 0.8053 \times 5 = -5.117$

(*iii*) Adjustment of -5.117:

$(-5.117 \times 18) - 405$

$= -497 = y$

$\therefore$ Points are:

$x = 2200;\ y = -497$

No adjustment necessary.

$\therefore r$ is -0.982

APPENDIX IV

Normal Curve Areas

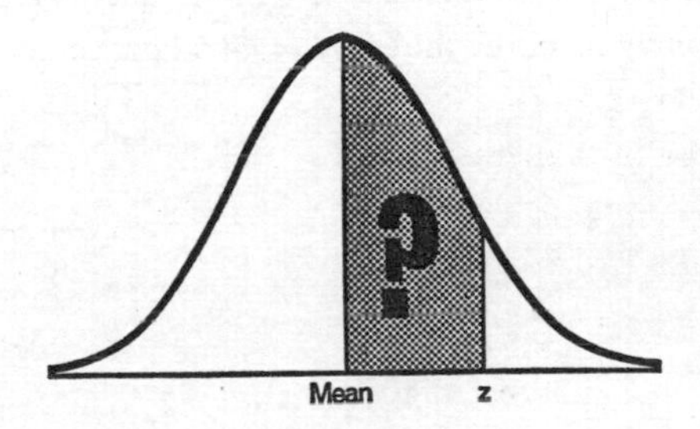

z is the distance the point lies from the mean measured in σ's, i.e.

$$z = \frac{Value - Mean}{\sigma}$$

(if z is minus, ignore sign).

z	*Area*	*z*	*Area*
0.0	0.0000	1.6	0.4452
0.1	0.0398	1.7	0.4554
0.2	0.0793	1.8	0.4641
0.3	0.1179	1.9	0.4713
0.4	0.1554	2.0	0.4772
0.5	0.1915	2.1	0.4821
0.6	0.2257	2.2	0.4861
0.7	0.2580	2.3	0.4893
0.8	0.2881	2.4	0.4918
0.9	0.3159	2.5	0.4938
1.0	0.3413	2.6	0.4953
1.1	0.3643	2.7	0.4965
1.2	0.3849	2.8	0.4974
1.3	0.4032	2.9	0.4981
1.4	0.4192	3.0	0.4987
1.5	0.4332		

EXAMPLE

If a distribution has a mean of 30 and a σ of 5, what area lies under the curve between the mean and 38?

$$\text{Now } z = \frac{38 - 30}{5} = 1.6$$

The table shows that when $z = 1.6$ the area is 0.4452.

$\therefore$ Area lying under the curve = 44.52 per cent.

APPENDIX V

(a) *Selected* t, χ^2 *and* e^{-x} *Values*

t and χ^2 *Cut-off Values*

Degrees of freedom	*Level of confidence*			
	t (two-tail)		χ^2	
	0·95	0·99	0·95	0·99
1	12.706	63.657	3.841	6.635
2	4.303	9.925	5.991	9.210
3	3.182	5.841	7.815	11.345
4	2.776	4.604	9.488	13.277
5	2.571	4.032	11.070	15.086
6	2.447	3.707	12.592	16.812
7	2.365	3.499	14.067	18.475
8	2.306	3.355	15.507	20.090
9	2.262	3.250	16.919	21.666
10	2.228	3.169	18.307	23.209
11	2.201	3.106	19.675	24.725
12	2.179	3.055	21.026	26.217
13	2.160	3.012	22.362	27.688
14	2.145	2.977	23.685	29.141
15	2.131	2.947	24.996	30.578
16	2.120	2.921	26.296	32.000
17	2.110	2.898	27.587	33.409
18	2.101	2.878	28.869	34.805
19	2.093	2.861	30.144	36.191
20	2.086	2.845	31.410	37.566
21	2.080	2.831	32.671	38.932
22	2.074	2.819	33.924	40.289
23	2.069	2.807	35.172	41.638
24	2.064	2.797	36.415	42.980
25	2.060	2.787	37.652	44.314
26	2.056	2.779	38.885	45.642
27	2.052	2.771	40.113	46.963
28	2.048	2.763	41.337	48.278
29	2.045	2.756	42.557	49.588
30	2.042	2.750	43.773	50.892

EXAMPLE 5 degrees of freedom, 0.95 level of confidence.

t interpretation: 95 per cent of all sample means of a sample size of 6 (one more than the degrees of freedom) fall within **2.571** standard errors of the population mean.

χ^2 interpretation: If χ^2 exceeds **11.070** and there are 5 degrees of freedom then one can say with 95 per cent confidence the difference between the actual and the expected results cannot be due solely to chance.

(b) Selected Negative Exponential Values

x	e^{-x}	x	e^{-x}
0.1	0.9048	3.1	0.0450
0.2	0.8187	3.2	0.0408
0.3	0.7408	3.3	0.0369
0.4	0.6703	3.4	0.0334
0.5	0.6065	3.5	0.0302
0.6	0.5488	3.6	0.0273
0.7	0.4966	3.7	0.0247
0.8	0.4493	3.8	0.0224
0.9	0.4066	3.9	0.0202
1.0	0.3679	4.0	0.0183
1.1	0.3329	4.1	0.0166
1.2	0.3012	4.2	0.0150
1.3	0.2725	4.3	0.0136
1.4	0.2466	4.4	0.0123
1.5	0.2231	4.5	0.0111
1.6	0.2019	4.6	0.0100
1.7	0.1827	4.7	0.00910
1.8	0.1653	4.8	0.00823
1.9	0.1496	4.9	0.00745
2.0	0.1353	5.0	0.00674
2.1	0.1225	5.1	0 00610
2.2	0.1108	5.2	0.00552
2.3	0.1003	5.3	0.00499
2.4	0.0907	5.4	0.00452
2.5	0.0821	5.5	0.00409
2.6	0.0743	5.6	0.00370
2.7	0.0672	5.7	0.00335
2.8	0.0608	5.8	0.00303
2.9	0.0550	5.9	0.00274
3.0	0.0498	6.0	0.00248

EXAMPLE $e^{-2.4} = 0.0907$

APPENDIX VI

Examination Technique

To pass any examination you must:

1. Have the knowledge.
2. Convince the examiner you have the knowledge.
3. Convince him within the time allowed.

In the book so far we have considered the first of these only. Success in the other two respects will be much more assured if you apply the examination hints given below.

1. Answer the question. Apart from ignorance, *failure to answer the question is undoubtedly the greatest bar to success.* No matter how often students are told, they always seem to be guilty of this fault. If you are asked for a frequency polygon, ***don't*** give a frequency curve; if asked to give the features of the mean, ***don't*** detail the steps for computing it. You can write a hundred pages of brilliant exposition, but if it's not in answer to the set question you will be given no more marks than if it had been a paragraph of utter drivel. To ensure you answer the question:

(*a*) *Read the question carefully.*
(*b*) *Decide what the examiner wants.*
(*c*) *Underline the nub of the question.*
(*d*) *Do just what the examiner asks.*
(*e*) *Keep returning to the question as you work on the answer.*

2. Put your ideas in logical order. It is quicker, more accurate and gives a greater impression of competence if you follow a predetermined logical path instead of jumping about from place to place as ideas come to you.

3. Maximise the points you make. Examiners are more impressed by a solid mass of points than an unending development of one solitary idea—no matter how sophisticated and exhaustive. Do not allow yourself to become bogged down with your favourite hobby-horse.

4. Allocate your time. Question marks often bear a close relationship to the time needed for an appropriate answer. Consequently the time spent on a question should be in proportion to the marks. Divide the total exam marks into the total exam time (less planning time) to obtain a "minutes per mark" figure, and allow that many minutes per mark of each individual question.

5. Attempt all questions asked for. Always remember that the first 50 per cent of the marks for any question is the easier to earn. Unless you are working in complete ignorance, you will always earn more marks per minute while answering a new question than while continuing to answer one that is more than half done. So you can earn many more marks by half-completing two answers than by completing either one individually.

6. Don't show your ignorance. Concentrate on displaying your knowledge—not your ignorance. There is almost always one question you need to attempt and are not happy about. In answer to such a question put down all you *do* know—and then devote the unused time to improving some other answer. Certainly you will not get full marks by doing this, but nor will you if you fill your page with nonsense. By spending the saved time on another answer you will at least be gaining the odd mark or so.

7. If time runs out. What should you do if you find time is running out? The following are the recommended tactics:

(*a*) If it is a mathematical answer, do not bother to work out the figures. Show the examiner by means of your layout that you know what steps need to be taken and which pieces of data are applicable. He is very much more concerned with this than with your ability to calculate.

(*b*) If it is an essay answer, put down your answer in the form of notes. It is surprising what a large percentage of the question marks can be obtained by a dozen terse, relevant notes.

(*c*) Make sure that every question and question part has some answer—no matter how short—that summarises the key elements.

(*d*) Don't worry. Shortage of time is more often a sign of knowing too much than too little.

8. Avoid panic, but welcome "nerves." "Nerves" are a great aid in examinations. Being nervous enables one to work at a much more concentrated pitch for a longer time without fatigue. Panic, on the other hand, destroys one's judgment. To avoid panic:

(*a*) Know your subject (this is your best "panic-killer").

(*b*) Give yourself a generous time allowance to read the paper. Quick starters are usually poor performers.

(*c*) Take two or three deep breaths.

(*d*) Concentrate simply on maximising your marks. Leave considerations of passing or failing until after.

(*e*) Answer the easiest questions first—it helps to build confidence.

(*f*) Do not let first impressions of the paper upset you. Given a few minutes, it is amazing what one's subconscious will throw up. Moreover it is often only the unfamiliar presentation of data that makes a statistical question look difficult: once you have looked carefully at it, it often shows itself to be quite simple.

SUGGESTED ANSWERS TO PROGRESS TESTS

Progress Test 1

1. (*a*) Or kill them more quickly. Or send them home unfit. A shorter stay in hospital does not of itself mean a quicker cure.

(*b*) The average driver covers more miles on numerous short trips near to home than he does on journeys outside the 5-mile range. Under such circumstances, therefore, it is not surprising that more accidents occur within 5 miles of the driver's home than outside this range. Long journeys are not then necessarily safer—they are just less common.

(*c*) Let us assume that *X* is black coffee. Now it may well have been that the 10 per cent of the drivers involved in an accident had drunk black coffee because they had been feeling drowsy, and that, in fact, their accident was due to this drowsiness. Clearly, then, the black coffee was not a contributory cause of their accident. Indeed, the 1 per cent who took coffee and did not have an accident might well have done so if they had gone without—and if 10 000 were *not* involved in an accident and only 100 were, then since 1 per cent of 10 000 is 100 this means that *not* taking black coffee could have resulted in double the number of accidents.

Progress Test 2

1. 280
500
641
800
900

3121 Ans. = 3100 tonnes

NOTE: Since three figures are exact hundreds, this implies that some figures are being rounded to the nearest 100. It is assumed, therefore, that 500, 800 and 900 are approximations, so the answer is approximated to the nearest 100 tons.

2. $1200 \times 112 \times 4 = 537\,600$.

Since the lowest number of significant figures in the rounded figures used is two (i.e. 1200 people) the answer can only contain two significant figures.

$\therefore$ Total approximate weight of potatoes bought in a year = 540 000 kg.

NOTE: The 4, being an exact number, is not a rounded number and therefore is excluded from the inspection for the number with the lowest number of significant figures.

3. (*a*) 21.388 ± 0.056. (*b*) 21.332 ± 0.056.

Progress Test 3

1. *Comments:*

(*a*) This is not a random sample. Indeed, there is every probability of bias since there is a tendency for people living in the same neighbourhood to be physically similar. Thus, if one newsagent was just outside an old persons' home it would be quite possible for the observer to record a preponderance of white-haired and frail readers.

(*b*) The *buyer* of a magazine is not necessarily the *reader*. It is quite possible that some housewives will pick up their husbands' regular order while they are out shopping.

(*c*) Physical "facts" collected in this way would not be very factual. Height and weight would need to be estimated and complexion would have to be judged subjectively.

The observers' results, then, might well conjure up in the mind of the reader a picture substantially different from the reality.

2. (*a*) Cluster sampling. (Random sampling is impossible as no sample frame exists here.)

(*b*) Multi-stage sampling and interviewing. (Interviewing will be necessary to obtain this type of personal information. The travelling involved as a result will render a random sample too expensive.)

(*c*) Stratified random sampling. (The number of West End, city, suburban and country cinemas can be found from official statistics.)

(*d*) Quota sampling.

Progress Test 4

1. *Changes in passenger journeys and receipts on British Railways,* 1958–61

(*a*) Passenger journeys

Type of fare	1958		1961		*Change**	
	No. of journeys (millions)	%	*No. of journeys (millions)*	%	*No. of journeys (millions)*	%
Full fare	351	32.2	273	26.6	−78	−5.6
Reduced fare	426	39.1	435	42.5	+9	+3.4
Season ticket	313	28.7	317	30.9	+4	+2.2
Total	1090	100	1025	100	−65	−†

(*b*) Receipts

Type of fare	1958		1961		*Change**	
	£ *millions*	%	£ *millions*	%	£ *millions*	%
Full fare	74	53.8	79.7	50.7	+5.7	−3.1
Reduced fare	46.1	33.5	54.1	34.4	+8	+0.9
Season ticket	17.5‡	12.7	23.4	14.9	+5.9	+2.2
Total	137.6	100	157.2	100	+19.6	−†

* Since the table is to bring out changes between the years, a section actually detailing the changes is a logical inclusion.

† Note that this total must come to zero, since the fares which increase as a *percentage of the total* must exactly balance the fares that decrease.

‡ It is a favourite examiner's technique to leave "holes" in the data for a table. The student must, of course, fill in the "holes" by logical deduction.

NOTES: The main decision in this question is whether to have the years as separate tables with passenger journeys and receipts side by side, or vice versa as shown in the answer. As the question asked for *changes* to be brought out, the layout adopted seemed preferable, the alternative layout being more appropriate to a table illustrating the *relationship between journeys and receipts* within the three-fare structure.

It was also tempting to consider using as alternative derived statistics the receipts per passenger journey (by dividing each receipt figure by its corresponding number of passenger journeys). However, valuable as these figures might be, they do not highlight *changes in the original data,* which is what the question seems to be aimed at.

2. *Criticisms of table:*

(*a*) No title.

(*b*) No source stated.

(*c*) No units given in "Weight of metal" column.

(*d*) (*i*) What are "foundry hours"?

(*ii*) Since all weights of castings are covered by the first three lines of the table, what can "Others" (line 4) refer to?

(*iii*) Does "Up to 4 kg" ("Up to 10 kg") include or exclude a casting weighing 4 (10) kg?

(All these are examples of ambiguity.)

(*e*) Since "Up to 10 kg" includes "Up to 4 kg" it would seem that double-counting is occurring.

(*f*) The "foundry hours" do not add up to 2000. What is this latter figure, then? If this total is relevant to the table, what is missing from the main body of it?

Summary. A confused table which, at best, tells little or nothing and, at worst, could mislead.

Progress Test 5

1. *See* Fig. 45.

NOTE: To answer the question it was necessary to use a double scale, and in order to bring out the relationship in the most effective manner the two scales were chosen so that the two curves occupied the same part of the graph.

As a result of such a choice of scales it becomes obvious from the graph that the two curves are almost identical, the main difference being that the rainfall curve *precedes* the profit curve by a year. This suggests that profits are closely related to the *previous* year's rainfall. (In statistics, when one curve follows another the time difference is termed *lag*. Finding cases of lag are often useful since it means that the future value of one variable can be closely estimated from the current value of another.)

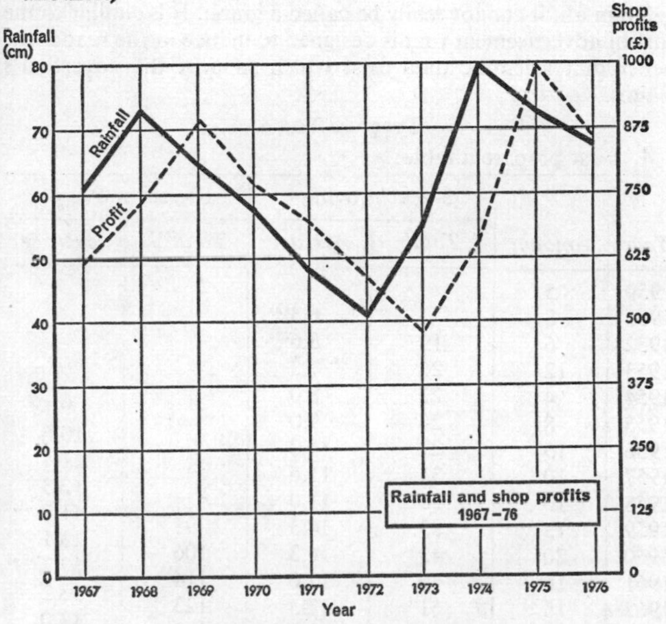

FIG. 45 *Rainfall and shop profits.*

2. (*a*) This graph has the following *serious* faults:

(*i*) The title is not clear. What is "improvement" and how is it measured?

(*ii*) The vertical axis shows neither heading nor units (it is presumed to be some sort of "improvement" scale—if, in fact, there can be such a scale). Moreover, it may not even start at zero.

(*iii*) The horizontal scale is clearly time, but no units are given at all.

(*iv*) The impression given is one of startling improvement. It is very much doubted if the figures on which the graph is based would support this impression (if there are any figures!).

(*v*) No source is stated.

(*b*) On the other hand:

(*i*) The curve is very distinct.

(*ii*) The graph is not overcrowded with curves.

(*iii*) The independent variable is correctly shown along the horizontal axis.

All in all, it cannot really be called a graph. It is similar to the sort of advertisement that is designed to induce in the reader the belief that scientific data exist which support the advertiser's claims.

Progress Test 6

1. First prepare a table:

		3-year moving–		10-year moving–	
Year	*Amount*	*Total*	*Average*	*Total*[1]	*Average*
1950	5				
1951	8		$6.\dot{3}$[2]		
1952	6	19	$8.\dot{6}$[3]		
1953	12	26	$7.\dot{3}$		
1954	4	22	8.0		9.1[4]
1955	8	24	9.0		10.6
1956	15	27	11.0		11.4
1957	10	33	$11.\dot{6}$		12.3
1958	10	35	11.0		11.7
1959	13	33	$14.\dot{3}$	91	13.1
1960	20	43	$16.\dot{3}$	106	13.2
1961	16	49	17.0	114	13.2
1962	15	51	$12.\dot{3}$	123	13.0
1963	6	37	13.0	117	13.2
1964	18	39	11.0	131	13.3
1965	9	33	14.0	132	13.1
1966	15	42	$10.\dot{6}$	132	13.7
1967	8	32	$11.\dot{6}$	130	13.8
1968	12	35	$11.\dot{3}$	132	14.6
1969	14	34	$14.\dot{6}$	133	14.8
1970	18	44	18.0	131	
1971	22	54	$18.\dot{6}$	137	
1972	16	56	$17.\dot{3}$	138	
1973	14	52	16.6	146	
1974	20	50		148	

Now *see* Fig. 46.

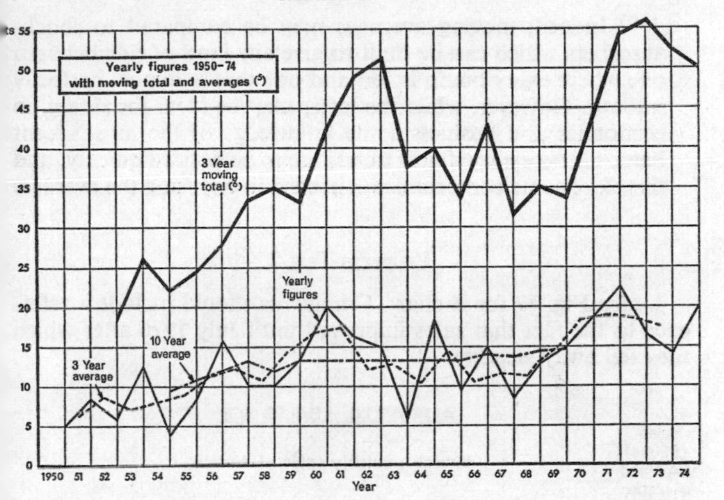

FIG. 46 *Yearly figures* 1950–74 *with moving total and averages.*

Difference between 10-year and 3-year moving averages: The 10-year moving average smooths out the fluctuations far more than the 3-year average—in fact, the 10-year average is nearly a straight line. However, the 3-year average is more sensitive to changes and signals a new trend sooner than the 10-year which tends to lag behind.(7)

NOTES: (1) To calculate a 10-year moving average it is first necessary to calculate a 10-year moving total.

(2) This average is located at the mid-point of its 3-year period.

(3) A dot over a decimal digit means that the decimal is recurring.

(4) This average is located at the mid-point of its 10-year period.

(5) Totals on this graph are plotted at the *end* of the period to which they apply, and averages at the *mid-point* (*see* V, 10(*b*)).

(6) A moving total of 3-years is really too short a period to give a curve of any significance. It has been included simply in order to give the student practice.

(7) Indeed, moving averages may be compared to shock-absorbers which can be built to give any kind of ride between one where every bump is felt and one that is almost perfectly smooth. However, while the latter may be ideal for riding, in economics and business some knowledge of the most recent bumps is necessary if new trends are to be noticed quickly, and therefore some compromise is needed in choosing the average.

Progress Test 7

1. *See* Fig. 47 for Z chart. Comments should include a reference to the fact that sales improved until July 1976 after which they fell away steadily.

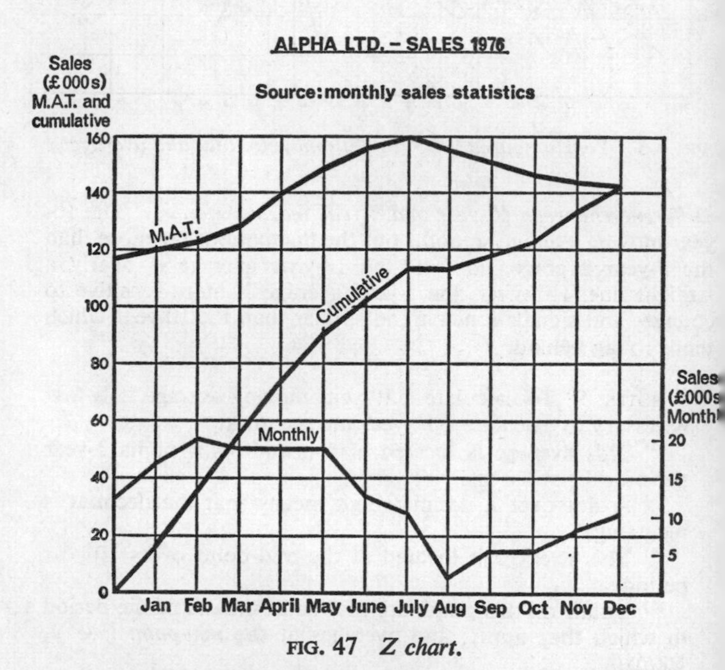

FIG. 47 *Z chart.*

2. *See* Fig. 48. This curve shows the extent to which the sharing of net output between establishments diverges from

Establishments			Net output		
No.	%	Cum. %	£000's	%	Cum. %
48	22½	22½	1406	3½	3½
42	19½	42	2263	6	9½
38	18	60	3699	9½	19
26*	12	72	3152*	8	27
21	10	82	2836	7½	34½
16	7	89	5032	13	47½
23	11	100	20385	52½	100
214			38773		

* *See* NOTE on page 266.

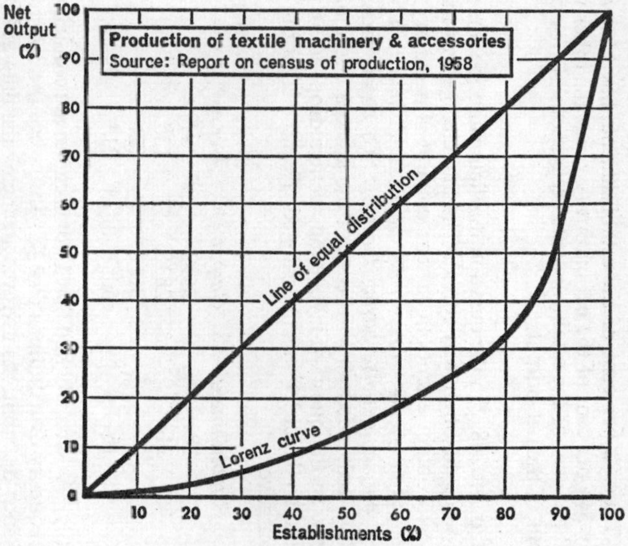

Production of textile machinery & accessories
Source: Report on census of production, 1958

Net output (%)
Line of equal distribution
Lorenz curve
Establishments (%)

FIG. 48 *Lorenz curve of production of textile machinery and accessories.*

equality. Thus 82 per cent of the establishments have between them only 34½ per cent of the net output—or, put the other way round, 18 per cent of the establishments are responsible for over 65 per cent of the net output.

NOTE TO FIG. 48. *In the question this figure and the one following were the other way round. The interchange was made because to obtain a smooth Lorenz curve it is necessary for *the order of the figure to run from those establishments having the least output to those having the most.* To check such an order it is only necessary to see that the *average* continually increases. In the question layout the average dropped at this point and so the order was re-arranged. Thus the layout in the question gave:

Establishments	*Output*	*Average*
38	3699	97½
21	2836	135
26	3152	121
16	5032	314½

Normally, students will find that Lorenz curves are constructed from frequency distributions (*see* IX) and, since the construction of such distributions involves arranging the data in order from the lowest to the highest, interchanges of this sort are rarely necessary.

Progress Test 9

1.

(*a*) *Age* (*years*)	(*b*) *Extractions*	(*c*) *Income p.a.* (£)
20–under 25	3–5†	Under 750§
25– „ 30	6–8	750– 1 249**
30– „ 35	9–11	1 250– 1 549††
35– „ 40	12–14	1 550– 1 849
40– „ 45	15–17	1 850– 2 149
45– „ 50	18–20	2 150– 2 449
50– „ 55	21–23	2 450– 2 749
55– „ 60	24–26	2 750– 3 249***
60– „ 65	27–29	3 250– 3 749
65– „ 70	30–32	3 750– 4 249
70– „ 75	33–34	4 250– 4 749
75– „ 80		4 750– 5 249

80– „ 85	Class limits "6–8":	5 250– 5 749
85– „ 90	5½ to 8½‡	5 750– 6 249
90– „ 95	extractions	6 250– 6 749
95– „ 100		6 750– 8 249
100 and over		8 250– 9 749
		9 750–13 249
Class limits of "25–under 30": 25 years exactly up to, but not including, 30 years*		13 250–17 749
		17 750–27 249
		27 250 and over§
		Class "750–1249": £749½ to £1249½‡‡

NOTES: * It should always be remembered that ages are usually given as at the last birthday, i.e. rounded down. Any distribution using ages should be constructed with this in mind. In fact, students are warned that official figures often reflect this by showing *stated* limits as 20–24; 25–29, etc., and so on.

† A class interval of 3 is not usually recommended, but it was chosen here as a compromise. An interval of 2 would be so small that one might as well construct the distribution without any grouping at all—and so avoid the loss of information that inevitably accompanies grouping. On the other hand, an interval of 4 would result in only 7 or 8 classes.

With 3 as an interval, the distribution "looks" better if the stated lower class limits are multiples of 3.

‡ The data is discrete—you cannot have half an extraction—so mathematical limits extending to half a unit on each side of the stated limits are used.

§ These classes are open-ended. It is assumed that very few full-time employed adults have incomes below £750 p.a. or over £27 250 p.a.

** This first group of classes has been chosen so that the round hundreds lie symmetrically throughout the class, as it is assumed that there will be a tendency for incomes to be set at the round hundred level. This arrangement is in compliance with suggestion (*d*), IX, **13** (note, however, that incomes set at the round £50 level will unfortunately result in some undesirable clustering at the beginning of classes).

†† Classes are unequal for reason given in IX, **11**.

*** Henceforth limits which ensure that round £1000s and £500s lie symmetrically throughout the class are selected.

‡‡ Clearly, if the data collected was recorded to the nearest £ then the *true* limits extend £½ out from the stated limits.

2. *Kilometres recorded by 120 salesmen in the course of one week*

Kilometres	Frequency (*f*)
390–under 410	4
410– " 430	17
430– " 450	33
450– " 470	31
470– " 490	20
490– " 510	13
510– " 530	2
	120

Progress Test 10

2. Set out the distribution:

Class	*f*	*Cumulative f*
0–under 10	12	12
10– " 25	25	37
25– " 40	51	88
40– " 50	48	136
50– " 60	46	182
60– " 80	54	236
80 and over	8	244

After laying out this distribution the first thing to decide, in view of the unequal class intervals, is what interval should be chosen for the histogram as "normal." Since three of the classes have intervals of 10, this interval has been chosen. This means that the other classes need adjustments to their frequencies in accordance with X, 2. These adjustments are:

Class	*Divide f by:*	*Adjusted f*
10–under 25	1.5	16.66
25– " 40	1.5	34
60– " 80	2	27
80– " 100	2	4

We can now construct the required graphs: *see* Fig. 49.

NOTES

* In plotting the adjusted frequencies the original frequencies do not show on the graph. Under these circumstances it is normal to write the original frequencies over the rectangles.

† Note that in the construction of ogives unequal class intervals do not lead to the sort of special adjustments needed when constructing histograms.

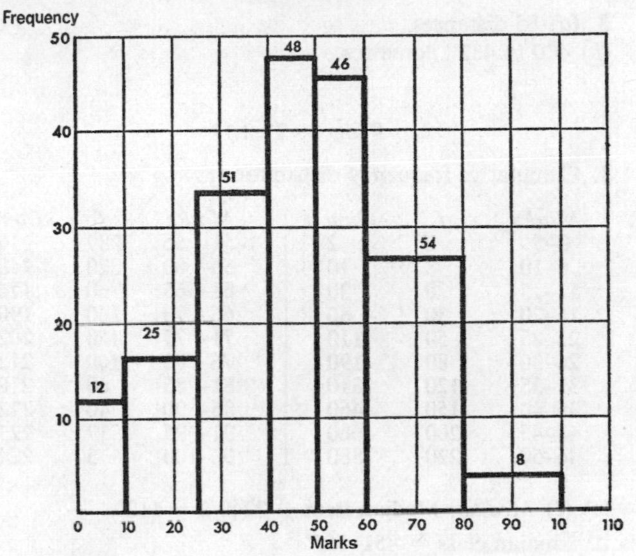

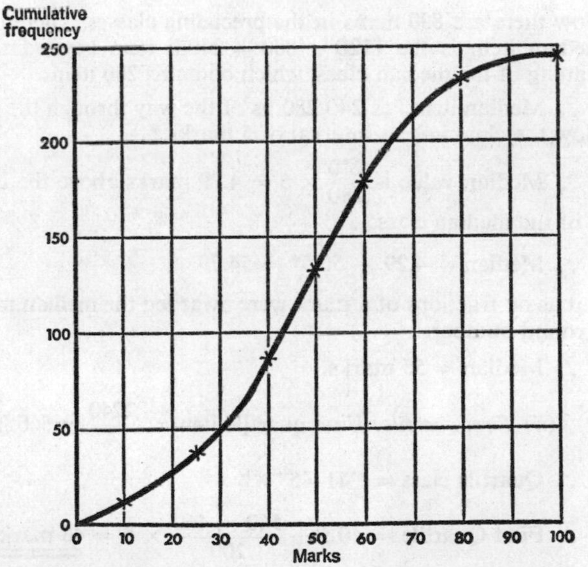

FIG. 49 *(a) Histogram of marks (first examination).**
(b) Ogive of marks (first examination).†

3. (a) 85 distances.
(b) 400 to 432 kilometres.

Progress Test 11

1. Cumulative frequency distribution:

Marks	f	Cum. f	Marks	f	Cum. f
0–5	2	2	51– 55	280	1160
6–10	8	10	56– 60	320	1480
11–15	20	30	61– 65	260	1740
16–20	30	60	66– 70	160	1900
21–25	50	110	71– 75	120	2020
26–30	80	190	76– 80	100	2120
31–35	120	310	81– 85	60	2180
36–40	150	460	86– 90	40	2220
41–45	200	660	91– 95	17	2237
46–50	220	880	96–100	3	2240

(a) (i) *Median:* Median item = 2240/2 = 1120.

∴ Median class = "51–55."

Now there are 880 items in the preceding classes. Therefore the median item is the 1120 − 880 = 240th item beyond the beginning of the median class, which contains 280 items.

∴ Median item lies 240/280ths of the way through the class. Now the class has an interval of 5 marks.*

∴ Median value is $\frac{240}{280} \times 5 = 4.29$ marks above the bottom of the median class.

∴ Median = 4.29 + 50.5* = 54.79

But as no fractions of a mark were awarded the median must be a round number.

∴ Median = 55 marks.

(ii) *First quartile:* First quartile item = $\frac{2240}{4}$ = 560th

∴ Quartile class = "41–45"

∴ First Quartile = $40.5 + \frac{560 - 460}{200} \times 5$ = 43 marks.

(*iii*) *Third quartile:* Third quartile item

$$= \frac{2240}{4} \times 3 = 1680\text{th}$$

$\therefore$ Quartile class = "61–65"

$$\therefore \text{ Third quartile} = 60.5 + \frac{1680 - 1480}{260} \times 5 = 64.35$$

i.e. rounded to the nearest unit = 64 marks.

(*iv*) *Sixth decile:* Sixth decile item $= \frac{2240}{10} \times 6 = 1344\text{th}$

$\therefore$ Sixth decile class = "56–60"

$$\therefore \text{ Sixth decile} = 55.5 + \frac{1344 - 1160}{320} \times 5 = 58.37$$

i.e. to the nearest unit = 58 marks.

(*v*) *Forty-second percentile:* Forty-second percentile item

$$= \frac{2240}{100} \times 42 = 940.8$$

i.e. since there can be no fraction of an item, 941st item.

$\therefore$ 42nd percentile class = "51–55"

$$\therefore \text{ 42nd percentile} = 50.5 + \frac{941 - 880}{280} \times 5 = 51.59$$

i.e. to nearest unit = 52 marks.

(*b*) If the examining body wished to pass only one-third of the candidates, the cut-off mark would need to be the mark obtained by the candidate one-third from the top, i.e. the $2240 - \frac{1}{3} \times 2240 = 1493$rd candidate.

This candidate's marks fall in the "61–65" class and can be estimated as $60.5 + \frac{1493 - 1480}{260} \times 5 = 60.75$

This means that the pass mark must be set at 61 marks.

NOTE: *Since no fractions of mark are given, the data is discrete and so the class limits are mathematical limits and are extended, therefore, a half mark either side of the stated limits, e.g. class "51–55" is considered to have limits of "50.5–55.5." This means that (*i*) the class interval is 5, and (*ii*) the class begins at 50.5.

2. 40th percentile = 445 kilometres.

(*a*) 500 km lies at the 93rd percentile.

(*b*) 450 km lies at the 45th percentile.

Progress Test 12

1. (*a*) Ungrouped frequency distribution of the data:

No. of passengers		*f*	*Cumulative f*
135	////	4	4
136	~~////~~ ///	8	12
137	~~////~~	5	17
138	///	3	20
		$\Sigma f = 20$	

(*i*) As these figures are very simple the short-cut method of calculating the mean is no quicker than the "direct method," so the latter method is used here:

No. (*x*)	*f*	*fx**
135	4	540
136	8	1088
137	5	685
138	3	414
	$n = 20$	$\Sigma fx = 2727$

$$\therefore \text{ Mean } (\bar{x}) = \frac{2727}{20} = 136.35 \text{ passengers.}$$

(*ii*) The most frequently recurring value is 136.
∴ Mode = 136 passengers.

(*iii*) The middle items of this array are the 10th and 11th. The cumulative frequency distribution shows that the values of both these items are the same, i.e. 136.
∴ Median = 136 passengers.

(*b*) If the last item was 35 instead of 135 then Σfx would be 100 less—i.e. 2627.

$$\therefore \text{ Mean } (\bar{x}) = \frac{2627}{20} = 131.35 \text{ passengers.}$$

This correction of the last item recorded simply means that one of the early items in the array has a value less than before, but this makes no difference to the values of the 10th and 11th items. Therefore the median remains unaltered. Similarly the mode remains unaltered since 136 still occurs most frequently.

The revised figures show that the mean can be greatly distorted by an extreme value. The revised mean now stands well below *any of the other 19 values in the distribution.* On the other hand the median and mode remain unaltered and this demonstrates that these two averages are not affected by extreme values.

NOTES ON ANSWER

* If a value is multiplied by the number of times it occurs (i.e. "*fx*"), the product is clearly equal to the figure obtained if all the occurrences of that value were added—e.g. $4 \times 135 = 135 + 135 + 135 + 135$. Therefore Σfx = the sum of all the values in the distribution.

2(*a*). *School A:*

Chosen class*: "105 to under 115."

I.Q. (x)	f	d	fd	Cum. f
75–under 85	15	−3	−45	15
85– " 95	25	−2	−50	40
95– " 105	40	−1	−40	80
105– " 115	108	0	0	188
115– " 125	92	+1	+92	280
125– " 135**	20	+2	+40	300
	$\Sigma f = 300$		$\Sigma fd = -3$	

Mean I.Q.*:*

$$\text{Now } \bar{x} = \text{Mid-point of chosen class} + \left(\frac{\Sigma fd}{\Sigma f} \times \text{Class interval}\right)$$

$$\therefore \bar{x} = 110 + \left(\frac{-3}{300} \times 10\right) = 110 - 0.1 = \underline{\underline{109.9 \text{ I.Q. marks.}}}$$

Median I.Q.*:* Median item = 300/2 = 150th item.

∴ Median class is "105–under 115"

Now the median item is 150 − 80 = 70th item in a class of 108 items.

∴ The median item lies 70/108ths into the class "105–under 115."

And since the class interval is 10 units, then 70/108ths of the interval $= \frac{70}{108} \times 10 = 6.5$ units.

∴ Median lies 6.5 units above the bottom of the "105–under 115" class, i.e. $105 + 6.5 =$ 111.5 I.Q. marks.

School B:

Similarly, mean = 95.3 I.Q. marks

median = 93.3 I.Q. marks.

NOTES: * *See page 243.*

** For the purpose of computing the mean an open-ended class is assumed to be the same size as the adjoining class (IX, 12).

(*b*)

School	*Mean* I.Q.	*No. of pupils*	*Total* I.Q. *points in school*
A	109.9	300	32 970
B	95.3	250	23 825
C	106.0	450	47 700
		Σw 1000	$\Sigma xw =$ 104 495

∴ Weighted average (mean I.Q. of all pupils in the three schools) $= \frac{104\,495}{1000} = 104.495$

$\simeq$ 104.5 I.Q. marks.

3. (*a*) Cumulative frequency distribution:

Length of wait (min)	*f*	*Cumulative f*	
0	50	50	
Under 0.5	210	260	
0.5–under 1	340	600	(First quartile class)
1– „ 2	200	800	(Median class)
2– „ 3	110	910	
3– „ 5	170	1080	(Third quartile class)
5– „ 10	140	1220	
10 and over	80	1300	

The middle 50 per cent of the customers are those who lie between the first and third quartiles (*see* XI, 9).

$$\text{First quartile item} = \frac{1300}{4} = 325\text{th item}$$

$\therefore$ First quartile

$$= 0.5 + \frac{325 - 260}{340} \times 0.5 = 0.5 + 0.096 = \underline{\underline{0.596}}$$

$$\text{Third quartile item} = \frac{1300}{4} \times 3 = 975\text{th item}$$

$$\therefore \text{Third quartile} = 3 + \frac{975 - 910}{170} \times 2 = 3 + 0.765 = \underline{\underline{3.765}}$$

$\therefore$ The middle 50 per cent of the customers wait between 0.596 and 3.765 minutes, i.e. between 36 seconds and 3 minutes 46 seconds.

(*b*) (*i*) Chosen class: 1–under 2 minutes

Length of wait (min)	*Mid-point*	*f*	*d*	*fd*
0	0	50	−1.5	−75
Over 0–under 0.5	0.25	210	−1.25	−262.5
0.5– „ 1	0.75	340	−0.75	−255
1– „ 2	1.5	200	0	0
2– „ 3	2.5	110	+1	+110
3– „ 5	4	170	+2.5	+425
5– „ 10	7.5	140	+6	+840
10 and over	12.5	80	+11	+880
		1300		+1662.5

Now since the formula for the mean of an unequal class interval distribution is:

$$\bar{x} = \text{Mid-point of chosen class} + \frac{\Sigma fd}{\Sigma f}$$

$$\text{Then } \bar{x} = 1.5 + \frac{1662.5}{1300} = \underline{\underline{2.78 \text{ minutes.}}}$$

(*ii*) From the cumulative frequency distribution drawn up in (*a*) it is possible to determine the median.

$$\text{Median item} = 1300/2 = 650\text{th}$$

$$\therefore \text{Median} = 1 + \frac{650 - 600}{200} \times 1 = 1 + 0.25 = \underline{\underline{1.25 \text{ minutes.}}}$$

This means that half the customers wait 1.25 minutes or less and the other half 1.25 minutes or more. Now the chances are equal as to which half you will be in,* and therefore you will have a 50/50 chance of being at the check point in 1.25 minutes or less. Since you wish to be there in 2 minutes, the odds are definitely in your favour.†

NOTES: * Look at it this way: *all* customers must fall into one half or the other. You are one of the customers. Since there is nothing to make you fall in one half more than another, the chance of being in the first half is the same as the chance of being in the second half.

† Students are sometimes puzzled as to why, if your mean wait is 2.78 minutes, you will have a good chance of reaching the check-out point in under 2 minutes. The reason is this:

The 2.78 minutes is the mean waiting time over a large number of visits. Now on some of these visits the waiting time was over 10 minutes. As was pointed out in this chapter, extreme values tend to distort the mean, and consequently these excessive waits make the mean waiting time longer than the majority of actual waiting times (even with the mean at 2.78 minutes, one wait of 10 minutes would need to be balanced by virtually three visits with no waiting time at all if the mean were to stay unchanged).

4. (*a*)

Machine	*No. of minutes per article*	*No. of articles per hour*
A	2	60/2 = 30
B	3	60/3 = 20
C	5	60/5 = 12
D	6	60/6 = 10
	Total articles per hour =	72

(*b*)

Machine	*Production* In 1st and 2nd hour	In 3rd hour	*Total*
A	0	30	30
B	40	20	60
C	24	0	24
D	20	10	30
			144

∴ 144 articles were produced in the 3 hours.

∴ Average number of articles per hour
= 144/3 = 48 articles.

5. (*a*) *The arithmetic mean.* The few high wages will result in this mean being the highest of the averages. Although such extreme wage payments would be unlikely to "distort" the mean seriously, their effect would result in the average looking larger than the figures in the whole of the distribution would warrant.

(*b*) *The mode.* In this sort of distribution the most frequently occurring wage is inevitably near the bottom, and choice of this average would result in a figure which made no allowance for the few high wages that some members of the foremen's union would be earning.

(*c*) *The median.* The median wage indicates the wage of the foreman who is "half way up the ladder." Half the foremen do better and half do worse. So you have a 50/50 chance of earning the median wage or better.

6. A combined mean is found by multiplying the number of items in each group by the respective means of each group; adding the products; and finally dividing by total number of items. As a formula this can be written so:

$$\text{Combined mean} = \frac{n_1\bar{x}_1 + n_2\bar{x}_2 + n_3\bar{x}_3}{n_1 + n_2 + n_3}$$

Progress Test 13

1. *School A.*

NOTE: In Progress Test 12, Question 2(*a*), $\bar{x}$ was found to be 109.9 I.Q. marks.

Standard deviation:

I.Q. (*x*)	*f*	*d*	*fd*	*fd*²	*MP**	(*MP* − $\bar{x}$)	*f*(*MP* − $\bar{x}$)
75–under 85	15	−3	−45	135	80	29.9	448.5
85– „ 95	25	−2	−50	100	90	19.9	497.5
95– „ 105	40	−1	−40	40	100	9.9	396.0
105– „ 115	108	0	0	0	110	0.1	10.8
115– „ 125	92	+1	+92	92	120	10.1	929.2
125 and over	20	+2	+40	80	130	20.1	402.0
Σ	300		−3	447			2 684.0

$$\text{Since } \sigma = \sqrt{\frac{\Sigma fd^2}{f} - \left(\frac{\Sigma fd}{f}\right)^2} \times \text{Class interval}$$

*Class mid-point.

then $\sigma = \sqrt{\frac{447}{300} - \left(\frac{-3}{300}\right)^2} \times 10$

$= \sqrt{\frac{447}{300} - \frac{1}{10\,000}} \times 10$

$= \sqrt{1.49 - 0.0001} \times 10 = \underline{\underline{12.2 \text{ I.Q. units}}}$

Mean deviation:

Mean deviation

$$= \frac{\Sigma f(MP - \bar{x})}{\Sigma f} = \frac{2684}{300} = 8.95 \simeq \underline{\underline{9 \text{ I.Q. marks}}}$$

Variability:

Coefficient of variation

$$= \frac{\sigma}{\bar{x}} \times 100 = \frac{12.2}{109.9} \times 100 = \underline{\underline{11.1}}$$

School B.

Similarly, $\sigma = \underline{\underline{11.2 \text{ I.Q. marks}}}$, mean deviation $= \underline{\underline{9\frac{1}{2} \text{ I.Q. marks}}}$,

and coefficient of variation $= \underline{\underline{11.75}}$.

Therefore the variability of the I.Q.s in School B is slightly greater than that of School A.

2. $Q_1 = 28.84$ pence and $Q_3 = 42.52$ pence.

Therefore quartile deviation $= \frac{42.52 - 28.84}{2}$

$= \underline{\underline{6.84 \text{ pence}}}$

3. Standard deviation = 6.96

4. $Sk = \frac{3(\text{Mean} - \text{Median})}{\sigma}$

$\therefore\ Sk = \frac{3(454.5 - 452.4)}{27} = \frac{3 \times 2.1}{27} = +0.233$

This indicates that the direction of skew is *positive*, and that the degree of skew is 0.233.

5.

Sales (£000)	*f*	*d*	*fd*	fd^2	*Cum. f*
Under 10	25	−1	−25	25	25
10–under 20	18	0	0	0	43
20– „ 30	8	+1	+8	8	51
30– „ 40	3	+2	+6	12	54
40 and over	1	+3	+3	9	55
	55		−8	54	

$$\therefore\ \bar{x} = 15 + \frac{-8}{55} \times 10 = 13.55, \text{ i.e. mean is } \underline{\underline{£13\,550}}$$

Median item is the 28th.

$$\therefore\ \text{Median} = 10 + \frac{3}{18} \times 10 = 11.67, \text{ i.e. median is } \underline{\underline{£11\,670}}$$

$$\sigma = \sqrt{\frac{54}{55} - \left(\frac{-8}{55}\right)^2} \times 10 = 9.8,$$

i.e. standard deviation is $\underline{\underline{£9800}}$

$$\therefore\ \text{Skewness} = \frac{3(13\,550 - 11\,670)}{9800} = \underline{\underline{+0.575}}$$

6.

$$Sk = \frac{3(20 - 22)}{10} = \frac{3 \times -2}{10} = -0.6$$

$\therefore$ Coefficient of skewness = $\underline{\underline{-0.6}}$

(i.e. it is *negatively* skewed).

Progress Test 14

1. *See* Fig. 50. Note that I.Q. is the independent variable and is therefore plotted on the horizontal axis.

(*a*) I.Q. 130: estimated mark 80.
(*b*) Mark 77: estimated I.Q. 127.

2. *See* Fig. 51. It is not possible to draw a line of best fit on this graph. Therefore the variables are not related.

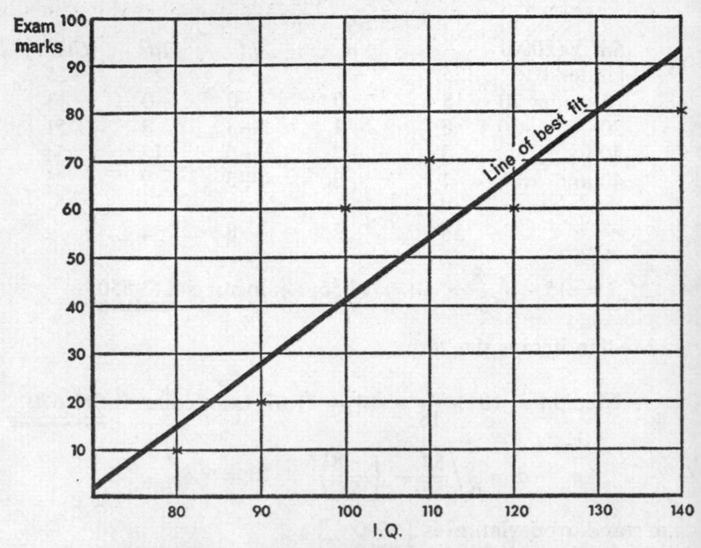

FIG. 50 *Scattergraph of data in Question* 1.

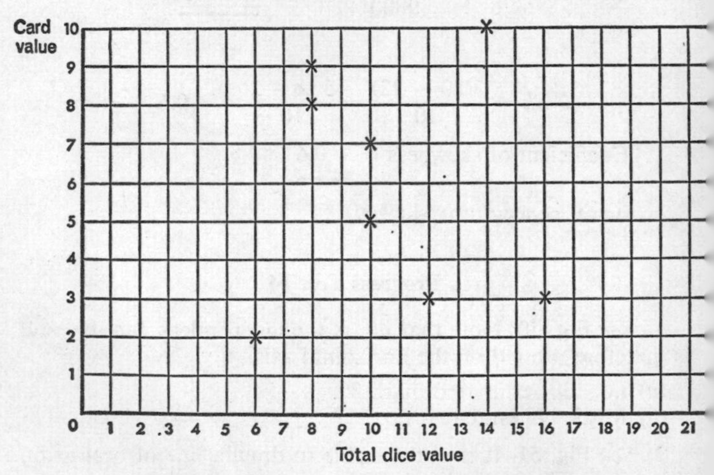

FIG. 51 *Scattergraph of data in Question* 2.

Progress Test 15

1. (*a*) *Regression line of examination marks on* I.Q.

$y = a + bx$, where y = examination marks and x = I.Q.

Now $\Sigma y = an + b\Sigma x$ and $\Sigma xy = a\Sigma x + b\Sigma x^2$.

	x^*	y^*	xy	x^2
	11	7	77	121
	10	6	60	100
	14	8	112	196
	12	6	72	144
	8	1	8	64
	9	2	18	81
Σ's:	64	30	347	706

and $n = 6$

$\therefore\ 30 = 6a + 64b$ (*i*)

and $347 = 64a + 706b$ (*ii*)

(*i*) $\times 32 : 960 = 192a + 2048b$

(*ii*) $\times 3 : 1041 = 192a + 2118b$

$\therefore\ 81 = 70b$

$\therefore\ b = 1.16$

Substitution of 1.16 for b in (*i*) gives $30 = 6a + 64 \times 1.16$

$= 6a + 74.2$

$\therefore\ a = -7.4$

$\therefore\ y = -7.4 + 1.16x$

Now, the x's were scaled down to 1/10 their original values for this calculation. Therefore, if their full values were incorporated in the equation, the constant b would need to be reduced by 1/10 to compensate, i.e. 0.116. However, the y's also were scaled down to 1/10, so the existing equation will give a y value only 1/10 of the true. To adjust for this, the whole of the right-hand side of the equation (including the new b figure) needs to be multiplied by 10. Hence the true equation is:

$$y = -74 + 1.16x$$

If a candidate had an I.Q. of 130, the best estimate of his mark would be:

$$y = -74 + 1.16 \times 130 = \underline{\underline{77 \text{ marks}}}$$

(*b*) *Regression line of* I.Q. *on marks*

$x = a + by$†

Now $\Sigma x = an + b\Sigma y$
and $\Sigma xy = a\Sigma y - b\Sigma y^2$
$\Sigma y^2 = 7^2 + 6^2 + 8^2 + 6^2 + 1^2 + 2^2 = 190$
$\therefore\ 64 = 6a + 30b$ (*i*)
and $347 = 30a + 190b$ (*ii*)
Solving ultimately gives $x = 73 + 0.675y$

Best estimate of the I.Q. of a candidate who obtained 77 marks:

$$x(\text{I.Q.}) = 73 + 0.675 \times 77 = \underline{\underline{125}}$$

See Fig. 52 for the superimposition of these lines on the scatter-graph in Fig. 50.

NOTE: * The "short" method is used here (*see* Method 6 in Appendix III), though since the adjustments are simple, at the end the whole of the regression line equation will be re-adjusted.

† The *a* and *b* in this equation are, of course, quite different from the *a* and *b* of the equation in part (*a*) of the answer (though the *x*'s and *y*'s relate to the same variable).

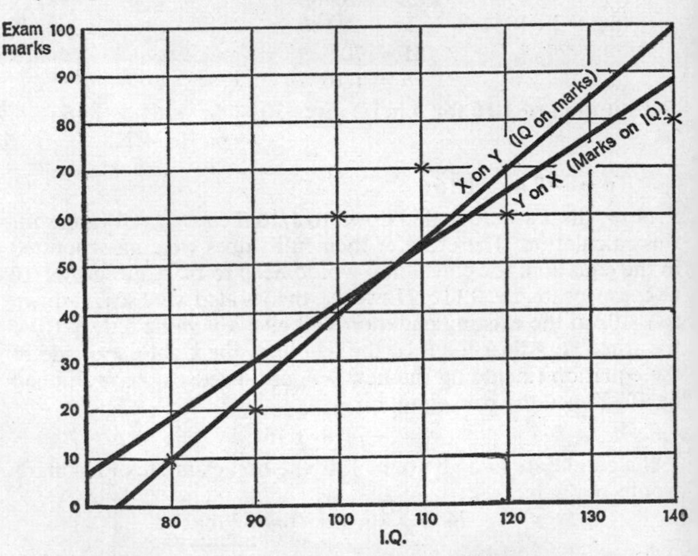

FIG. 52 *Regression lines of data in Question* 1, p. 111 *superimposed on scattergraph in Fig.* 50.

2. Let x = die value, and y = card value, where $n = 10$.

	x	y	x^2	y^2	xy
	8	8	64	64	64
	8	9	64	81	72
	14	10	196	100	140
	22	5	484	25	110
	22	8	484	64	176
	16	3	256	9	48
	12	3	144	9	36
	6	2	36	4	12
	10	7	100	49	70
	10	5	100	25	50
Σ's:	128	60	1928	430	778

(*a*) *Regression line of card value on die value*

Appropriate line equation: $y = a + bx$.

Simultaneous equations: $\Sigma y = an + b\Sigma x$

$$\Sigma xy = a\Sigma x + b\Sigma x^2$$

$\therefore\ 60 = 10a + 128b$ (*i*)

and $778 = 128a + 1928b$ (*ii*)

Solving gives $\underline{y = 5.56 + 0.0345x}$

$\therefore$ Estimate of card value "y" when die value "x" is:

(*i*) 4: $y = 5.56 + 0.0345 \times 4 = 5.70$*

(*ii*) 24: $y = 5.56 + 0.0345 \times 24 = 6.39$*

(*b*) *Regression line of die value on card value*

Appropriate line equation: $x = a + by$

Simultaneous equations: $\Sigma x = an + b\Sigma y$

$$\Sigma xy = a\Sigma y + b\Sigma y^2$$

$\therefore\ 128 = 10a + 60b$

and $778 = 60a + 430b$

Solving gives $\underline{x = 11.94 + 0.14y}$

∴ Estimate of die value "x" when card value "y" is:

(*i*) $1: x = 11.96 + 0.14 \times 1 \quad = \underline{\underline{12.1}}$*

(*ii*) $10: x = 11.96 + 0.14 \times 10 = \underline{\underline{13.36}}$*

* In these examples it is not possible, of course, for a card or die value to be other than a whole number and strictly speaking the estimates should be rounded to the nearest unit. However, the answer here is left in decimal form so that the difference between estimates (*i*) and (*ii*) can be shown.

See Fig. 53 for the superimposition of the regression lines on the scattergraph in Fig. 51.

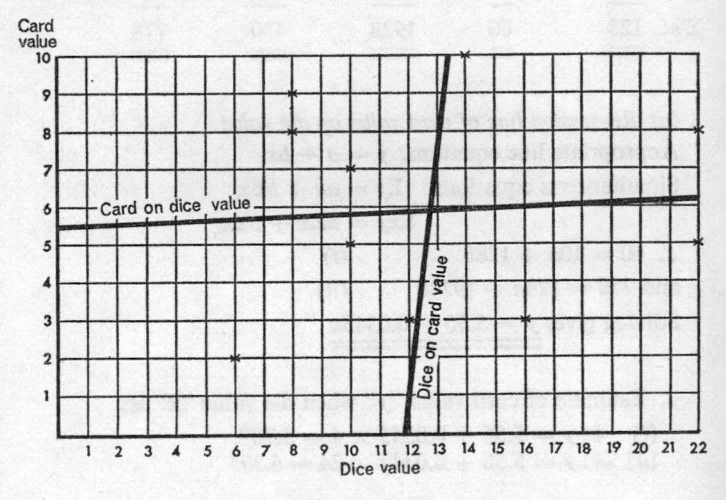

FIG. 53 *Regression lines of data in Question* 2, *superimposed on Fig.* 51.

(*c*) In (*a*) estimates of card values for the two *extreme* dice values were virtually the same—i.e. 5.70 and 6.39. Indeed, since decimal values are impossible in this case, rounding to the nearest unit does give identical estimates of 6. This only confirms a commonsense view of what is being said. If two variables are clearly unrelated (as must be the case when random card values and random dice values are recorded) then the value of one will be quite independent of the value of the other, and the best estimate of the former will be its mean value. Of course, such an estimate will be quite unreliable, but the error which results from using the mean will probably be smaller than that arising from the use of any other value. The best card value estimate, then, is the mean card value, i.e. 6.

This value of 6 is the best estimate of card value for all dice values. This means that on a graph, having dice values on the horizontal axis, the "line of best fit" is a *horizontal line* opposite "6" on the vertical axis. Considering now the estimation of *dice values*, using the same arguments as above it is clear that the best estimate is the *mean die value*, no matter what the card value might be. On our graph this will give a *vertical line* opposite 12.8 on the horizontal axis. This means that the "line of best fit" suitable for estimating a y value from an x value is quite a different line from that suitable for estimating an x value from a y value.

In our example the regression lines are not quite horizontal and vertical. As the variables are quite unrelated they ought to be, but chance factors governing the values actually recorded has resulted in a very slight associationship appearing to exist between card and dice values. Such false slight associationship commonly arises in statistics due to these chance factors—an absolutely correct result is, in fact, as improbable as obtaining 500 heads and 500 tails in a 1000 tosses of a coin.

Progress Test 16

1. (*a*) *Means*

x^*	y^*
11	7
10	6
14	8
12	6
8	1
9	2
64	30

$$\therefore \bar{x} = \frac{64}{6} = \underline{\underline{10.66}} \qquad \therefore \bar{y} = \frac{30}{6} = \underline{\underline{5}}$$

(*b*) *Standard deviations*

Note that σ_x will be calculated using an assumed mean of 11†. σ_y will be calculated using the direct method.

x	$(x-\bar{x})$	$(x-\bar{x})^2$	y	$(y-\bar{y})$	$(y-\bar{y})^2$
11	0	0	7	+2	4
10	−1	1	6	+1	1
14	+3	9	8	+3	9
12	+1	1	6	+1	1
8	−3	9	1	−4	16
9	−2	4	2	−3	9
	−2	24			40

$$\therefore \sigma_x = \sqrt{\frac{24}{6} - \left(\frac{-2}{6}\right)^2} = \sqrt{4 - \frac{1}{9}} = \sqrt{\frac{35}{9}} \qquad \therefore \sigma_y = \sqrt{\frac{40}{6}} = \sqrt{\frac{20}{3}}$$

(*c*) Σxy

x	y	xy
11	7	77
10	6	60
14	8	112
12	6	72
8	1	8
9	2	18
	$\Sigma xy =$	347

(*d*) *Applying r formula,*

$$r = \frac{347 - 6 \times 10.66 \times 5}{6 \times \sqrt{\frac{35}{9}} \times \sqrt{\frac{20}{3}}} = \frac{347 - 320}{6 \times \sqrt{\frac{35}{9} \times \frac{20}{3}}}$$

$$= \frac{27}{6 \times \sqrt{\frac{700}{27}}} = +0.885‡$$

NOTES: * It is quite permissible to scale down or up either or both series of figures in order to simplify calculations. Note that (*i*) the scaling of one series is completely independent of any scaling of the other series; (*ii*) since correlation is not concerned with absolute values—only degrees of change—no adjustments for such scaling need to be made later in the calculations.

† Note that it simplifies calculation to use an assumed mean even though the true mean is already known. The method used is Method 4 in Appendix III.

‡ Note that r is not expressed in any units.

2.

x	$(x - \bar{x})$*	$(x - \bar{x})^2$	y	$(y - \bar{y})$	$(y - \bar{y})^2$	xy
8	−4	16	8	+2	4	64
8	−4	16	9	+3	9	72
14	+2	4	10	+4	16	140
22	+10	100	5	−1	1	110
22	+10	100	8	+2	4	176
16	+4	16	3	−3	9	48
12	0	0	3	−3	9	36
6	−6	36	2	−4	16	12
10	−2	4	7	+1	1	70
10	−2	4	5	−1	1	50
128	+8	296	60		70	778
Σx	$\Sigma(x - \bar{x})$	$\Sigma(x - \bar{x})^2$	Σy		$\Sigma(y - \bar{y})^2$	Σxy

x series

$$\bar{x} = \frac{128}{10} = 12.8$$

$$\sigma_x = \sqrt{\frac{296}{10} - \left(\frac{8}{10}\right)^2} = \sqrt{\frac{724}{25}} = \frac{1}{5}\sqrt{724}$$

y series

$$\bar{y} = \frac{60}{10} = 6$$

$$\sigma_y = \sqrt{\frac{70}{10}} = \sqrt{7}$$

$$\therefore r = \frac{778 - 10 \times 12.8 \times 6}{10 \times \frac{1}{5}\sqrt{724} \times \sqrt{7}} = \frac{778 - 768}{2 \times \sqrt{724 \times 7}}$$

$$= +0.07$$

The type of correlation is *very low positive*, so low as to indicate that the variables are virtually uncorrelated.‡

Since there is no true correlation between the value of a thrown die and the value of a drawn card, such slight correlation that there is must obviously be *spurious*.§

NOTES: Note by careful layout all the Σ's can be computed from a single table.

* Assumed mean = 12.

‡ *See* 4.

§ *See* 6.

3.

Person	I.Q. *ranking*	*Exam ranking*
A	3	2
B	4	3 equal (i.e. 3.5*)
C	1	1
D	2	3 equal (i.e. 3.5*)
E	6	6
F	5	5

$$r' = 1 - \frac{6\,\Sigma d^2}{n(n^2 - 1)}$$

Computation of Σd^2

I.Q.	*Exam*	d	d^2
3	2	1	1
4	3.5	0.5	0.25
1	1	0	0
2	3.5	1.5	2.25
6	6	0	0
5	5	0	0
		$\Sigma d^2 =$	3.5

$$\therefore\ r' = 1 - \frac{6 \times 3.5}{6 \times (6^2 - 1)} = \underline{\underline{+0.9}}$$

In Question 1, r was 0.885. The difference arises through the fact that when ranking is used instead of the full set of figures there is some loss of information. This loss is reflected in the correlation values.†

NOTES: * Where there is a tie, the ranking given is the average of the ranks shared. In this case, persons B and D share the 3rd and 4th ranks. They are therefore both given the average of 3 and 4, i.e. 3.5.

† The more accurate correlation value is, of course, the one computed from the full set of figures—in this case 0.885.

4.

Means:

$$\bar{x} = (5 + 75)/2 = \underline{40}$$

$$\bar{y} = (17 + 19)/2 = \underline{18}$$

Standard deviations:

x	$(x - \bar{x})$	$(x - \bar{x})^2$	y	$(y - \bar{y})$	$(y - \bar{y})^2$
5	−35	1225	17	−1	1
75	+35	1225	19	+1	1
		2450			2

$$\therefore \sigma_x = \sqrt{\frac{2450}{2}} = \underline{35} \qquad \therefore \sigma_y = \sqrt{\frac{2}{2}} = \underline{1}$$

Σxy:

$$\Sigma xy = 5 \times 17 + 75 \times 19 = 85 + 1425 = \underline{1510}$$

Apply formula:

$$r = \frac{\Sigma xy - n\bar{x}\bar{y}}{n\sigma_x\sigma_y}$$

$$= \frac{1510 - 2 \times 40 \times 18}{2 \times 35 \times 1} = \frac{1510 - 1440}{70} = \underline{\underline{+1}}$$

This answer is rather unexpected since the original figures do not, on the face of it, appear to be perfectly correlated. However, the reason for this answer is easily explained, for if there are only two points on a scattergraph the "line of best fit" is simply a line joining the points. Now in the case where all the points lie on the "line of best fit" there is perfect correlation. Thus there is always perfect correlation where there are only two pairs of figures—as in this case.

The conclusion that follows from this is that the interpretation of r depends, among other things, upon the number of pairs of figures. Two pairs give perfect correlation, but each additional pair beyond this make it increasingly difficult to find a "line of best fit" that falls close to all the points. Thus the value of r will tend to diminish as more pairs of figures are added. (Note, however, this change in the value of r is only of importance where the number of pairs of figures is small.)

5. No, an error need not have occurred. The following explanation could quite easily account for the result:

Hours of study is not the only factor influencing examination success. Maturity and intellectual ability could both be vital factors also. Now it is well known that people with high maturity and/or high intellectual ability need to study for fewer hours than their less fortunate colleagues in order to reach the same level of knowledge. If now the examination favours students with high maturity and/or intellectual ability then the success of such students will be greater than their colleagues, despite the extra hours of studying by the latter. Thus examination marks and hours of study would have a high negative correlation.

NOTE ON ANSWER: This example illustrates the difference between relationship and cause. Although fewer hours of study are related to examination success it is obviously quite wrong to argue that such success is *due to* the few hours' study, and that therefore the less one studies the better one's chances of success.

Progress Test 17

1. (*a*) (*i*) $\frac{1}{2} \times \frac{1}{2} = \frac{1}{4}$

(*ii*) $\frac{1}{6} \times \frac{1}{6} = \frac{1}{36}$

(*b*) $\frac{1}{6} + \frac{1}{6} = \frac{1}{3}$

(*c*) The probability of drawing the ace of spades *or* clubs *or* hearts *or* diamonds $= \frac{1}{52} + \frac{1}{52} + \frac{1}{52} + \frac{1}{52} = \frac{1}{13}$

(*d*) String-events are H—T, T—H

∴ Probability $= (\frac{1}{2} \times \frac{1}{2}) + (\frac{1}{2} \times \frac{1}{2}) = \frac{1}{2}$

(*e*) String-events giving a total of 11: 5—6, 6—5

∴ Probability, therefore, of 11 $= (\frac{1}{6} \times \frac{1}{6}) + (\frac{1}{6} \times \frac{1}{6}) = \frac{1}{18}$[(1)]

(*f*) String-events giving total of 7: 1—6, 2—5, 3—4, 4—3, 5—2, 6—1

∴ Probability of 7 $= (\frac{1}{6} \times \frac{1}{6}) + (\frac{1}{6} \times \frac{1}{6}) + (\frac{1}{6} \times \frac{1}{6}) + (\frac{1}{6} \times \frac{1}{6}) + (\frac{1}{6} \times \frac{1}{6}) + (\frac{1}{6} \times \frac{1}{6}) = \frac{1}{6}$

(*g*) This is the "at least" situation, so the probability of backing at least one winner is 1 — probability of backing 0 winners. We need, then, 1 — probability of all horses losing

$$= 1 - (\tfrac{2}{3} \times \tfrac{3}{5} \times \tfrac{27}{40} \times \tfrac{5}{9}) = 0.85$$

NOTE: ([1]) It should be appreciated that the probability of throwing a 5 and a 6 is *not* the same as throwing two 6's. The reason is that there are two different string-events giving 5 and 6 (5—6 and 6—5) but only one of a double 6 (6—6).

2. (*a*) Probability both will be alive $= \frac{5}{8} \times \frac{5}{6} = \underline{\underline{\frac{25}{48}}}$

(*b*) Probability at least one of them will be alive = 1 — probability both will be dead $= 1 - (\frac{3}{8} \times \frac{1}{6}) = \underline{\underline{\frac{15}{16}}}$

(*c*) Probability only wife will be alive = Probability husband dead *and* wife alive $= \frac{3}{8} \times \frac{5}{6} = \underline{\underline{\frac{5}{16}}}$.

3. Probability that A will obtain a given contract $= \frac{10}{20} = 0.5$, that B will obtain a given contract $= \frac{6}{20} = 0.3$ and that neither will obtain a given contract $= \frac{4}{20} = 0.2$.

(*a*) Probability that A will obtain all three contracts

$$= 0.5 \times 0.5 \times 0.5 = \underline{\underline{0.125}}$$

(*b*) Probability that B will obtain at least one contract = 1 — probability that B obtains no contracts

$$= 1 - (0.7 \times 0.7 \times 0.7) = \underline{\underline{0.657}}$$

(*c*) Probability that 1st and 2nd contract, or 1st and 3rd contract, or 2nd and 3rd contract obtained by neither*

$$= (0.2 \times 0.2 \times 0.8) + (0.2 \times 0.8 \times 0.2) + (0.8 \times 0.2 \times 0.2) = \underline{\underline{0.096}}$$

(*d*) Probability that A obtains 1st contract, B 2nd contract and A 3rd contract $= 0.5 \times 0.3 \times 0.5 = \underline{\underline{0.075}}$.

NOTE * This assumes that the remaining contract *is* to A or B.

4. (*a*) Note that the probability of my being late is the probability of my being less than 5 minutes late + the probability of my being 5 or more minutes late (since this is the "or/exclusive" situation) = 0.2 + 0.7 = 0.9. Since in this question the probability of arriving is 1 (certain), we need to know what the missing 0.1 is. Clearly, here it is the probability of not being late—when, of course, no reprimand will be given. So:

String-event	*Probability*
Early–no reprimand	0.1 × 1 = 0.10
Less than 5 mins–no reprimand	0.7 × 0.6 = 0.42
5 mins or more–no reprimand	0.2 × 0.1 = 0.02

∴ Probability of no reprimand = $\underline{\underline{0.54}}$

(Note alternative calculation: Probability of no reprimand = 1 − probability of reprimand (as computed below).)

(*b*) Probability of reprimand:

String-event	*Probability*
Less than 5 mins–reprimand	$0.7 \times 0.4 = 0.28$
5 mins or more–reprimand	$0.2 \times 0.9 = 0.18$

$\therefore$ Probability of reprimand = 0.46*

$\therefore$ Probability of being less than 5 minutes late *given*

$$\text{reprimand} = P(\text{Less than 5 mins late} \mid \text{reprimand}) = \frac{0.28}{0.46} = 0.609$$

NOTE * Strictly speaking we should account for the third possible string-event—the "early-reprimand" event. However, since this comes to $0.1 \times 0 = 0$, it can be ignored in the layout.

5. First we must find the probability of an "unreasonable" bill. So we have:

String-event	*Probability*
Car unused by son–bill unreasonable	$0.7 \times 0.8 = 0.56$
Car used by son–bill unreasonable	$0.3 \times 1.0 = 0.30$

$\therefore$ Probability of unreasonable bill = 0.86

The best way to proceed now is to find the probability that my son did *not* use the car *given an unreasonable bill.* So:

$$P(\text{car unused by son} \mid \text{unreasonable bill}) = \frac{0.56}{0.86} = 0.651$$

Since the question is, what is the probability that my son used the car at all during the three months, we can say this is 1 − probability he never used it. The probability that he never used it in any of the three months, given that the bill was unreasonable in each month, is the probability that he never used it in the

1st month *and* never used it in the 2nd month *and* never used it in the 3rd month = 0.651 × 0.651 × 0.651 = 0.276.

∴ Probability my son used the car to some extent

$$= 1 - 0.276 = \underline{\underline{0.724}}$$

Progress Test 18

1. Here $n = 10$, $p = 0.1$ and $q = 1 - 0.1 = 0.9$

(*a*) (*i*) The binomial distribution formula in this case is:

$$P(x) = \binom{n}{x} p^x q^{n-x} = \frac{10!}{(10-x)!\,x!} \times 0.1^x \times 0.9^{10-x}$$

The layout, then, is as follows:

x (*Houses without bathrooms*	$P(x)$
0	$\frac{10!}{(10-0)! \times 0!} \times 0.1^0 \times 0.9^{10-0} = 0.3487$
1	$\frac{10!}{(10-1)! \times 1!} \times 0.1^1 \times 0.9^{10-1} = 0.3874$
2	$\frac{10!}{(10-2)! \times 2!} \times 0.1^2 \times 0.9^{10-2} = 0.1937$
3	$\frac{10!}{(10-3)! \times 3!} \times 0.1^3 \times 0.9^{10-3} = 0.0574$
4	$\frac{10!}{(10-4)! \times 4!} \times 0.1^4 \times 0.9^{10-4} = 0.0112$
5	$\frac{10!}{(10-5)! \times 5!} \times 0.1^5 \times 0.9^{10-5} = 0.0015$
6 or more	Balance = 0.0001
	$\underline{\underline{1.0000}}$

(*ii*) Here $a = np = 0.1 \times 10 = 1$

The Poisson distribution formula, therefore, is:

$$P(x) = e^{-a} \times \frac{a^x}{x!} = e^{-1} \times \frac{1^x}{x!} = \frac{e^{-1} \times 1^x}{x!}$$

Now since $e^{-1} = 0.3679$ (Appendix V) and 1^x always equals 1 no matter what x might be, then $e^{-1} \times 1^x = 0.3679$. So we have $P(x) = \frac{0.3679}{x!}$. The layout, then, is:

x (*Houses without bathrooms*)	P(x)
0	$\frac{0.3679}{0!} = 0.3679$
1	$\frac{0.3679}{1!} = 0.3679$
2	$\frac{0.3679}{2!} = 0.1840$
3	$\frac{0.3679}{3!} = 0.0613$
4	$\frac{0.3679}{4!} = 0.0153$
5	$\frac{0.3679}{5!} = 0.0031$
6 or more	Balance = 0.0005
	1.0000

(*b*) *See* Fig. 54.

(*c*) The probability that less than two houses in the sample will lack bathrooms is (using the binomial distribution) the probability that there will be exactly 0 houses without bathrooms + the probability there will be exactly 1 house

$$= 0.3487 + 0.3874 = \underline{\underline{0.7361}}$$

NOTE: Comparison of the two distributions reveals small but definite differences—though even these are under 7 per cent in the first four classes. However, as was observed when the question was set, the situation is a very borderline one for using a Poisson distribution. In more appropriate circumstances the differences would be very much smaller.

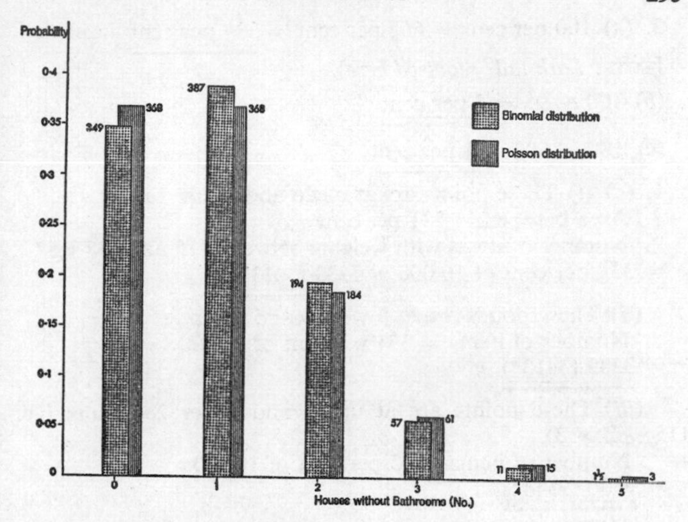

FIG. 54 *Binomial and Poisson distributions compared when $n = 10$ and $p = 0.1$.*

Progress Test 19

[*Spurious accuracy is allowed here to assist students' checking.*]

1. (*a*) $\frac{1}{2} \times 95$ per cent $= 47\frac{1}{2}$ per cent

(*b*) $\frac{1}{2} \times 99\frac{3}{4}$ per cent $= 49\frac{7}{8}$ per cent

(*c*) From Appendix IV, 49.53 per cent

2. (*a*) Mean to lower $2\sigma = 47\frac{1}{2}$ per cent
Mean to lower $1\sigma = 33\frac{1}{3}$ per cent
$\therefore$ Lower 1σ to lower $2\sigma = 47\frac{1}{2} - 33\frac{1}{3} = 14\frac{1}{6}$ per cent

NOTE: Actual is 13.6 per cent*.

(*b*) Lower 1σ to upper $2\sigma = 33\frac{1}{3} + 47\frac{1}{2} = 80\frac{5}{6}$ per cent

(*c*) Lower 2σ to upper $3\sigma = 47\frac{1}{2} + 49\frac{7}{8} = 97\frac{3}{8}$ per cent

(*d*) Lower 2.6σ to upper $2.6\sigma = 2 \times 49.53 \simeq 99$ per cent

3. (*a*) 100 per cent − $66\frac{2}{3}$ per cent = $33\frac{1}{3}$ per cent

[NOTE: *Both tails required here*]

(*b*) 100 − 95 = 5 per cent

(*c*) 100 − $99\frac{3}{4}$ = $\frac{1}{4}$ per cent

4. (*a*) (*i*) These points are at mean and upper 1σ
∴ Area between = $33\frac{1}{3}$ per cent
∴ Number of items with weights between 115 and 118 kg
= $33\frac{1}{3}$ per cent of 10 000 = 3333 (3413*)

(*ii*) These points are at lower 1σ and mean.
∴ Number of items = $33\frac{1}{3}$ per cent of 10 000
= 3333 (3413*)

(*iii*) These points are at upper and lower 2σ points (i.e. $115 \pm 2 \times 3$).
∴ Number of items = 95 per cent of 10 000
= 9500 (9544*)

(*iv*) These points are at upper and lower 3σ (i.e. $115 \pm 3 \times 3$).
∴ Number of items = $99\frac{3}{4}$ per cent of 10 000
= 9975 (9974*)

(*b*) 109 to 121 kg is $115 \pm 2 \times 3$, i.e. these weights lie at the 2σ points. Now, 95 per cent of all the distribution falls between these points. Therefore 19 out of 20 items will be between 109 and 121 kg, so any prediction that the weight of an item selected at random will lie between 109 and 121 kg can be made at a 95 per cent level of confidence.

NOTE: * This is the figure obtained if the table in Appendix IV is used.

Progress Test 20

1. (*i*) $\sigma_{\bar{x}} = \dfrac{15}{\sqrt{25}} = 3$ kg

(*a*) Mean estimate at 95 per cent level
$= 950 \pm 2 \times 3 =$ 944 to 956 kg

(*b*) Mean estimate at 99.75 per cent level
$= 950 \pm 3 \times 3 =$ 941 to 959 kg

(*ii*) $\sigma_{\bar{x}} = \dfrac{0.8}{\sqrt{100}} = 0.08$ cm

(*a*) Mean estimate at 95 per cent level

$= 1.82 \pm 2 \times 0.08 = \underline{\underline{1.66 \text{ cm to } 1.98 \text{ cm}}}$

(*b*) Mean estimate at 99.75 per cent level

$= 1.82 \pm 3 \times 0.08 = \underline{\underline{1.58 \text{ cm to } 2.06 \text{ cm}}}$

(*iii*) $\sigma_{\bar{x}} = \dfrac{0.8}{\sqrt{10\,000}} = 0.008$

(*a*) Mean estimate at 95 per cent level

$= 1.82 \pm 2 \times 0.008 = \underline{\underline{1.804 \text{ cm to } 1.836 \text{ cm}}}$

(*b*) Mean estimate at 99.75 per cent level

$= 1.82 \pm 3 \times 0.008 = \underline{\underline{1.796 \text{ cm to } 1.844 \text{ cm}}}$

NOTE: the narrower limits that follow the use of a larger sample size.

2. (*i*) $\sigma_p = \sqrt{\dfrac{0.61 \times 0.39}{100}} = 0.0488$

Proportion estimate 95 per cent level

$= 0.61 \pm 2 \times 0.0488 = 0.512$ to 0.708

$\simeq \underline{\underline{51.2 \text{ per cent to } 70.8 \text{ per cent males}}}$

(*ii*) $\sigma_p = \sqrt{\dfrac{0.61 \times 0.39}{10\,000}} = 0.00488$

Proportion estimate 95 per cent level

$= 0.61 \pm 2 \times 0.00488$

$\simeq \underline{\underline{60.0 \text{ per cent to } 62.0 \text{ per cent males}}}$

(*iii*) $\sigma_p = \sqrt{\dfrac{\frac{26}{49} \times \frac{23}{49}}{49}} = 0.071$

Proportion estimate 95 per cent level

$= 26/49 \pm 2 \times 0.071$

$\simeq \underline{\underline{38.8 \text{ per cent to } 67.2 \text{ per cent defectives}}}$

NOTE: that when estimating proportions a very much larger sample size is required than is needed when estimating means in order to give estimate limits which are close enough to be worth while.

3. If the estimate reads ± 2 cm at the 95 per cent level of confidence, then 2 cm $= 2 \times \sigma_{\bar{x}}$.

$\therefore \sigma_{\bar{x}} = 1$ cm

Since $\sigma_{\bar{x}} = \frac{\sigma_x}{\sqrt{n}}$, then $1 = \frac{\sigma_x}{\sqrt{100}} = \frac{\sigma_x}{10}$

$\therefore \sigma_x$ (i.e. σ of sample) $= 10$ cm

The investigator wants an estimate within $\frac{1}{2}$ cm at the 95 per cent level of confidence.

$\therefore \frac{1}{2}$ cm $= 2 \times \sigma_{\bar{x}}$ (i.e. the improved $\sigma_{\bar{x}}$)

$\therefore \sigma_{\bar{x}} = \frac{1}{4}$ cm

Again, since $\sigma_{\bar{x}} = \frac{\sigma_x}{\sqrt{n}}, \frac{1}{4} = \frac{10}{\sqrt{n}}$

$\therefore \sqrt{n} = 40$

$\therefore n = 1600$

$\therefore$ The sample size will need to be 1600.

4. If a 99 per cent confidence level is required, it entails finding how far either side the mean one has to measure in order to include 99 per cent of the total area beneath the normal curve or, in other words how far *one side* of the mean one has to measure to include 99/2 = 49.5 per cent of the total area. Looking up the table in Appendix IV, it can be seen that an area of 0.4953 is embraced by using 2.6σ. Therefore $\sigma_{\bar{x}}$ needs to be multiplied by 2.6 to give a confidence level of 99 per cent.

Progress Test 21

1. Null hypothesis: there is no contradiction between a population mean of 14 hours and a sample mean of 13 hours 20 minutes.

Test: Two-tail.

Now $\sigma_{\bar{x}} = \frac{\sigma_x}{\sqrt{n}} = \frac{3}{\sqrt{64}} = \frac{3}{8}$ hour.

If 14 hours is the population mean, then 95 per cent sample means will fall between $14 \pm 2 \times \frac{3}{8} = 13\frac{1}{4}$ and $14\frac{3}{4}$ hours.

The sample mean is $13\frac{1}{3}$ hours and therefore lies between these limits.

Therefore the null hypothesis cannot be rejected at the 95 per cent level of confidence, and the welfare officer's assertion is not disproved.

2. Null hypothesis: there is no contradiction between a population proportion of 0.6 and a sample proportion of 1410/2500 = 0.564.

Test: Two-tail (the candidate has *not* asserted that *at least* 60 per cent of the voters support him).

Now if $p = 0.6^*$, $q = 1 - 0.6 = 0.4$, and $n = 2500$, then:

$$\sigma_p = \sqrt{\frac{pq}{n}} = \sqrt{\frac{0.6 \times 0.4}{2500}} = 0.0098$$

If 0.6 is the population proportion, then $99\frac{3}{4}$ per cent sample proportions will fall between $0.6 \pm 3 \times 0.0098 = 0.5706$ and 0.6294.

The sample proportion is 0.564, which is *outside* these limits, hence the null hypothesis must be rejected at the $99\frac{3}{4}$ per cent level of confidence, i.e. there *is* a contradiction between the asserted population proportion and the sample proportion.

This means that the difference is significant and the population proportion is not 60 per cent.†

NOTES: * Remember: use the true population proportion for p where possible (*see* XXI, 9). Although the true proportion is not known here, it represents the true proportion until such time as it is proved untrue.

† Note how the size of the sample improves our discrimination. The candidate claimed in effect that 1500 out of every 2500 people supported him. The sample gave 1410: a mere 90 short, on a sample of perhaps only 3 per cent or 4 per cent of the total voters. Nevertheless, with a sample of 2500 we were able to say that his assertion was wrong—and say this, moreover, knowing we would only make an error of wrongly dismissing such a claim once in 400 times. (We are, of course, assuming the people sampled told the truth!)

3. If the two samples were taken from the same population, it would be equivalent to two samples taken from two populations having the same mean. Therefore, to test the expert's assertion, we need to find out whether the difference between the means is significant.

Now $\sigma_{\bar{x}_1} = \dfrac{2}{\sqrt{200}}$ and $\sigma_{\bar{x}_2} = \dfrac{2}{\sqrt{80}}$.

Since $\sigma_{(\bar{x}_1-\bar{x}_2)} = \sqrt{\sigma^2_{\bar{x}_1} + \sigma^2_{\bar{x}_2}}$, then

$$\sigma_{(\bar{x}_1-\bar{x}_2)} = \sqrt{\left(\frac{2}{\sqrt{200}}\right)^2 + \left(\frac{2}{\sqrt{80}}\right)^2} = \sqrt{\frac{4}{200}+\frac{4}{80}}$$
$$= \underline{\underline{0.264 \text{ kg}}}$$

Therefore 95 per cent of the differences between means can extend to $2 \times \sigma_{(\bar{x}_1-\bar{x}_2)} = 2 \times 0.264 = 0.528$ kg.

The actual difference is $20\frac{1}{2} - 20 = 0.5$ kg. This is within the allowed limits and there is, therefore, no evidence to disprove the expert's assertion that the two samples were taken from a single population.

4. This question calls for a one-tail test of significance of the difference between proportions (one-tail, since the question is not as to whether there is a difference between the fitness of the citizens of the two cities but as to whether there is a difference showing that citizens of A specifically are fitter than those of B).

Now the proportion of citizens passing in A is 96/200 = 0.48 and the proportion of citizens passing in B is 84/200 = 0.42.

$$\therefore \sigma_{p_A} = \sqrt{\frac{pq}{n}} = \sqrt{\frac{0.48 \times 0.52^*}{200}}$$

$$\text{and} \quad \sigma_{p_B} = \sqrt{\frac{0.42 \times 0.58^*}{200}}$$

Since $\sigma_{(p_1-p_2)} = \sqrt{\sigma_{p_A} + \sigma_{p_B}}$

$$\text{then } \sigma_{(p_1-p_2)} = \sqrt{\left(\sqrt{\frac{0.48 \times 0.52}{200}}\right)^2 + \left(\sqrt{\frac{0.42 \times 0.58}{200}}\right)^2}$$
$$= \sqrt{\frac{0.48 \times 0.52}{200} + \frac{0.42 \times 0.58}{200}} = 0.049\,66^{**}$$

Now a 95 per cent one-tail level of confidence is the same as a 90 per cent two-tail level of confidence. The table in Appendix IV shows that a z of 1.65 (interpolating) will give an area 0.45 of the curve, i.e. 90 per cent when doubled. So 95 per cent of the differences between proportions, where one given proportion is greater than the other, do not exceed 1.65 standard errors. Our 95 per cent confidence limit, then, is $1.65 \times 0.049\,66 = 0.0819$.

And the actual difference between the proportions is $0.48 - 0.42 = 0.06$. Since this difference is within the allowed limit it is not significant.

This means that the difference between the 96 fit citizens of A and the 84 fit citizens of B could have arisen by chance alone and the health official's claim that the citizens of A are fitter than those of B is not proven.†

NOTES: * It pays not to simplify these expressions, since at the next stage they are squared, and that means that one has only to remove the square root sign.

** Using a pooled proportion we would have in lieu:

$$\text{Pooled proportion} = \frac{96 + 84}{200 + 200} = \frac{180}{400} = 0.45$$

Since $\sigma_{(p_1-p_2)} = \sqrt{pq\left(\frac{1}{n_1}+\frac{1}{n_2}\right)}$, then,

$$\sigma_{(p_1-p_2)} = \sqrt{0.45 \times 0.55 \times \left(\frac{1}{200} + \frac{1}{200}\right)} = 0.049\,75$$

As can be seen, this adjustment has altered the standard error only very slightly.

† Note how a small sample in a proportion problem can mean that quite large differences can occur without the differences being significant (thus, in this case, a difference of 96 — 84 = 12 in a sample size of 200, i.e. 6 per cent, is not significant).

Progress Test 22

1. $\sigma_{\bar{x}} = \frac{\sigma}{\sqrt{n}} = \frac{0.120}{\sqrt{16}} = 0.030$ cm.

With 16 — 1 = 15 degrees of freedom and at the 95 per cent level of confidence $t = 2.131$ (*see* Appendix V).

$$\begin{aligned}\therefore \text{ Estimate of population mean} &= \text{Sample mean} \pm t \times \sigma_{\bar{x}} \\ &= 6.214 \pm 2.131 \times 0.030 \\ &= \underline{\underline{6.150 \text{ cm to } 6.278 \text{ cm}}}\end{aligned}$$

NOTE: Strictly speaking, as the sample is a small sample, we should adjust the sample standard deviation so that we can obtain a more accurate estimate of the population standard deviation. This adjustment is made as follows:

$$\text{Since } \sigma_{sample} = \sqrt{\frac{\Sigma(x - \bar{x})^2}{n}} = \frac{\sqrt{\Sigma(x - \bar{x})^2}}{\sqrt{n}}$$

then $\sigma_{sample} \times \sqrt{n} = \sqrt{\Sigma(x - \bar{x})^2}$

$$\therefore \sqrt{\frac{\Sigma(x - \bar{x})^2}{n - 1}} = \sigma_{sample} \times \frac{\sqrt{n}}{\sqrt{n - 1}} = \sigma_{sample} \times \frac{\sqrt{n}}{\sqrt{n - 1}}$$

= estimate of population standard deviation.

∴ Estimate of population standard deviation in this case

$$= \frac{0.120 \times \sqrt{16}}{\sqrt{16 - 1}} = 0.124.$$

Using this 0.124 in lieu of 0.120 results in the estimate of the population mean being 6.148 cm to 6.280 cm, i.e. a slight increase in the confidence interval.

2. Finding the standard deviation of the sample by the direct method we have:

x	$(x - \bar{x})$	$(x - \bar{x})^2$
122	3	9
133	14	196
102	−17	289
3)357		494
$\therefore \bar{x} = 119$ g		$\sigma = \sqrt{494/3}$

∴ Estimate of the population standard deviation

$$= \sqrt{\frac{494}{3 - 1}} = \sqrt{247} \text{ g}$$

$$\therefore \sigma_{\bar{x}} = \frac{\sqrt{247}}{\sqrt{3}} = \sqrt{82.33} = 9.07 \text{ g}$$

With 3 − 1 = 2 degrees of freedom and at the 95 per cent level of confidence $t = 4.303$ (*see* Appendix V).

∴ Estimate at the 95 per cent level of confidence of the mean weight of the apples in the barrel

$$= 119 \pm 4.303 \times 9.07 = \underline{\underline{79.97 \text{ g to } 158.03 \text{ g}}}$$

Now, since the number of apples in the barrel

$$= \frac{\text{Total weight of apples}}{\text{Average weight per apple}},$$

then estimate of number of apples in barrel

$$= \frac{45.5 \times 10^3}{158.03} \text{ to } \frac{45.5 \times 10^3}{79.97}$$

$$= \underline{\underline{288 \text{ to } 569}}$$

NOTE: (*i*) In this estimate the margin of error is so large that the estimate is virtually useless. This example demonstrates, therefore, the fact that a very small sample gives only a very small amount of information about the population.

(*ii*) In practice, of course, it would be quicker (as well as more accurate) to count all the apples in the barrel than to spend time selecting a genuine random sample, weighing the apples selected and then calculating the estimate!

3. Here we find the mean difference and then test if this differs significantly from the population mean of zero that would exist if, in fact, alcohol had no effect on the speed of response. Note that a two-tail test is called for since the test aims at detecting *any* difference—good or bad—and not just a difference in one direction.

Volunteer:	*A*	*B*	*C*	*D*	*E*	*Total*
Speed Factor:						
Before	12.5	10.2	11.7	9.4	12.1	
After	12.9	10.3	12.5	9.6	12.6	
Difference (x)	+0.4	+0.1	+0.8	+0.2	+0.5	+2.0
$(x - \bar{x})^2$	0	0.09	0.16	0.04	0.01	0.30

$$\bar{x} = \frac{+2.0}{5} = +0.4$$

Estimate of population standard deviation

$$= \sqrt{\frac{0.3}{5 - 1}} = 0.274.$$

$$\text{Standard error } (\sigma_{\bar{x}}) = \frac{0.274}{\sqrt{5}} = 0.123.$$

So sample mean of 0.4 lies $\frac{0.4}{0.123} = 3.252$ standard errors from zero.

Degrees of freedom $= 5 - 1 = 4$.

Tables show that at the 95 per cent level of confidence where there are 4 degrees of freedom $t = 2.776$.

Since the sample mean lies well beyond this limit the differences are highly significant. Alcohol does, therefore, very probably have a effect on the speed of response.

NOTE: If the student compares the *Before* figures with the Test B figures in **XXII 10** he will see that the former are exactly one-tenth the latter. The mean difference in the speed of response test, however, is only +0.4 while that in the I.Q. test is +5. So the former mean shows a relatively smaller increase than the latter, *yet this former mean is significant whereas the* I.Q. *mean is not*. The reason for this is that the *variability* in the alcohol test was much less than that in the I.Q. test (i.e. the coefficients of variation are

$$\sqrt{\frac{0.3}{5}} \div 0.4 = 0.612 \text{ and } \sqrt{\frac{124}{5}} \div 5 = 0.996 \text{ respectively})$$

and hence the mean increase, while lower overall, was more consistent. Clearly, the more consistent a series of increases the less likely it is that they could have arisen through chance.

Progress Test 23

1. (*a*) Contingency table (with expected results in cells in brackets):

Factory	*Quality rating*			*Total*
	Poor	*Medium*	*Good*	
X	26 (20)*	35 (36)	19 (24)	80
Y	24 (30)	55 (54)	41 (36)	120
Total	50	90	60	200

(*b*) χ^2 computation:

A	E	$A - E$	$(A - E)^2$	$(A - E)^2/E$
26	20	+6	36	1.800
35	36	−1	1	0.028
19	24	−5	25	1.042
24	30	−6	36	1.200
55	54	+1	1	0.019
41	36	+5	25	0.694

$$\therefore \chi^2 = 4.783$$

(*c*) Degrees of freedom = 2

(*d*) *Tests.* Since χ^2 at the 95 per cent level of confidence and 2 degrees of freedom = 5.991 (*see* Appendix V) and since our computed χ^2 is less than this, then there is no significant difference between the results of the two factories and we cannot therefore say the quality of work done by Y is different to X (i.e. the observed difference in the sample figures could have arisen by chance more often than once in twenty times).

NOTES: * Since a total of 50 ratings out of 200 were poor then in the 80 total ratings made in factory X we would expect $\frac{50}{200} \times 80 = 20$ poor ratings. Similarly for each other cell in the table.

2. The first step here is to find *a*. This is computed as follows:

Points per Game	*Games*	*Total Points*
0	12	0
1	54	54
2	66	132
3	90	270
4	80	320
5	64	320
6	34	204
	400	1300

Since 400 games resulted in a total of 1300 points then

$$u = \frac{1300}{400} = 3.25$$

∴ Using the curve formula the relative frequencies are:

Number of games scoring x points

$$= 4 \times 3.25 - (3.25 - x)^2 = 13 - (3.25 - x)^2$$

Applying this formula gives us the relative frequencies in the second column below. However, this column only totals 62.58 games. Since the actual distribution had 400 games, it is necessary to multiply each relative frequency by $\frac{400}{62.58} = 6.39$, in order to obtain the theoretical "curve" that must be fitted to the actual distribution. The resulting theoretical games are shown in the third column.

The rest of the problem simply involves setting the actual distribution alongside the theoretical one, computing χ^2, and then seeing if this value of χ^2 exceeds the cut-off value given by the tables.

Points	*Relative Frequency*	*Theoretical Games*	*Actual Games*	$\frac{(A-E)^2}{E}$
0	2.44	15.59	12	0.8267
1	7.94	50.74	54	0.2095
2	11.44	73.10	66	0.6896
3	12.94	82.69	90	0.6462
4	12.44	79.49	80	0.0033
5	9.94	63.52	64	0.0036
6	5.44	34.76	34	0.0166
	62.58	399.89*	400	2.3955

* Rounding error of 0.11.

In fitting this curve we used the *a* of the actual distribution and also the total number of games (400). Deducting, then, 2 from the number of classes of 7 to allow for these used figures gives us 5 degrees of freedom. And χ^2 at the 95 per cent level of confidence with 5 degrees of freedom is shown in the table in Appendix V as 11.070. So the differences between the theoretical and actual games is not significant at the 95 per cent level of confidence—i.e. there are no grounds for saying the curve does not fit the observed data.

Progress Test 24

1. $n = 400; p = \frac{35}{1000} = 0.035$

$\therefore\ np = 400 \times 0.035 = 14$

and $\sqrt{npq} = \sqrt{400 \times 0.035 \times 0.965} = \sqrt{13.51} = 3.67$

$\therefore$ 95 per cent upper control limit

$= 14 + 3.67 \times 2 = 21.34$

99 per cent upper control limit

$= 14 + 3.67 \times 2.6^* = 23.55$

* Appendix IV shows 2.6σ above the mean relates to 49.53 per cent of the area (i.e. 2.6σ either side the mean embraces 99 per cent of the total area).

Now see Fig. 55 for control chart and plottings of hourly sample figures.

COMMENTS: By the eighth hour the process was out of control. Re-setting was, therefore, called for.

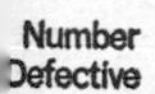

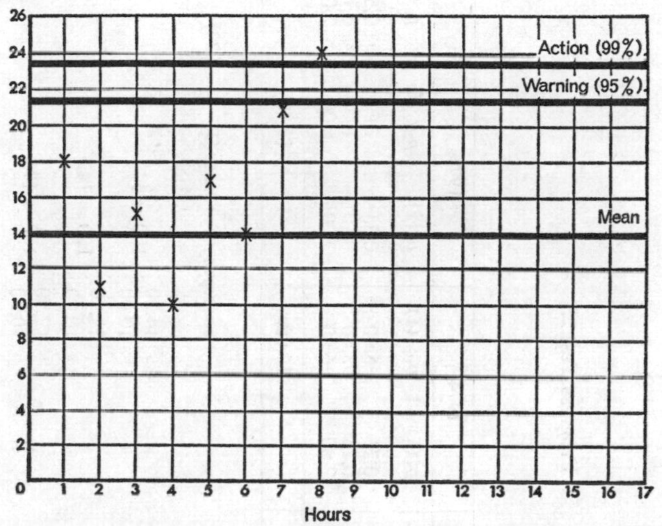

FIG. 55 *Number defective control chart.*

Progress Test 25

1. (*a*) (*i*) *Laspeyre price index:*

Item	1973	1973		1974		1975		1976	
	q_0	p_0 (£)	p_0q_0 (£)	p_1 (£)	p_1q_0 (£)	p_2 (£)	p_2q_0 (£)	p_3 (£)	p_3q_0 (£)
A	20	0.20	4.00	0.25	5.00	0.35	7.00	0.50	10.00
B	12	0.25	3.00	0.25	3.00	0.10	1.20	$0.12\frac{1}{2}$	1.50
C	3	1.00	3.00	2.00	6.00	2.00	6.00	2.00	6.00
Σ			10.00		14.00		14.20		17.50

Using the Laspeyre formula

Year	*Laspeyre price index* (1973 = 100)
1973	100
1974	$\frac{14.00}{10.00} \times 100 = 140$
1975	$\frac{14.20}{10.00} \times 100 = 142$
1976	$\frac{17.50}{10.00} \times 100 = 175$

1. (*a*) (*ii*) *Paasche price index:*

Item	1973	1974				1975				1976			
	p_0 (£)	p_1 (£)	q_1	p_1q_1 (£)	p_0q_1 (£)	p_2 (£)	q_2	p_2q_2 (£)	p_0q_2 (£)	p_3 (£)	q_3	p_3q_3 (£)	p_0q_3 (£)
A	0.20	0.25	24	6.00	4.80	0.35	20	7.00	4.00	0.50	18	9.00	3.60
B	0.25	0.25	16	4.00	4.00	0.10	20	2.00	5.00	$0.12\frac{1}{2}$	16	2.00	4.00
C	1.00	2.00	2	4.00	2.00	2.00	3	6.00	3.00	2.00	4	8.00	4.00
Σ				14.00	10.80			15.00	12.00			19.00	11.60

Using the Paasche formula

Year	*Paasche price index* (1973 = 100)
1973	100
1974	$\frac{14.00}{10.80} \times 100 = 130$
1975	$\frac{15.00}{12.00} \times 100 = 125$
1976	$\frac{19.00}{11.60} \times 100 = 164$

1. *(a)* *(iii)* ***Weighted average of price relatives:***

Basic data			1974			1975			1976		
Item	1973 p_0 (£)	*Weight*	p_1 (£)	$\frac{p_1}{p_0} \times 100$	*Price relative × Weight*	p_2 (£)	$\frac{p_2}{p_0} \times 100$	*Price relative × Weight*	p_3 (£)	$\frac{p_3}{p_0} \times 100$	*Price relative × Weight*
A	0.20	5	0.25	125	625	0.35	175	875	0.50	250	1250
B	0.25	3	0.25	100	300	0.10	40	120	0.12½	50	150
C	1.00	2	2.00	200	400	2.00	200	400	2.00	200	400
Σ		10			1325			1395			1800

Year	*Weighted average of price relatives* (1973 = 100)
1973	100
1974	$\frac{1325}{10} = 132.5$
1975	$\frac{1395}{10} = 139.5$
1976	$\frac{1800}{10} = 180$

Progress Test 26

1. (*a*) Since both series cover the same time period, space can be saved by laying out both series in the same table:

Year	*d**	d^2	*A*	*Ad*	*B*	*Bd*
1968	−4	16	25	−100	200	−800
1969	−3	9	31	−93	196	−588
1970	−2	4	36	−72	194	−388
1971	−1	1	44	−44	190	−190
1972	0	0	48	0	189	0
1973	+1	1	53	+53	185	+185
1974	+2	4	60	+120	184	+368
1975	+3	9	62	+186	181	+543
1976	+4	16	67	+268	180	+720
Σ's =		60	426	+318	1699	−150

Apply formulae for $y = a + bd$ equation.

A series

$$a = \bar{A} = \frac{\Sigma A}{n} = \frac{426}{9} = 47.3$$

$$b = \frac{\Sigma Ad}{\Sigma d^2} = \frac{+318}{60} = +5.3$$

∴ Population A (000's)

$$= 47.3 + 5.3d$$

B series

$$a = \bar{B} = \frac{\Sigma B}{n} = \frac{1699}{9} = 188.8$$

$$b = \frac{\Sigma Bd}{\Sigma d^2} = \frac{-150}{60} = -2.5$$

∴ Population B (000's)

$$= 188.8 - 2.5d\dagger$$

(*b*) When the populations are equal, then population A = population B

$$\text{i.e. } 47.3 + 5.3d = 188.8 - 2.5d$$
$$\therefore\ 5.3d + 2.5d = 188.8 - 47.3$$
$$\therefore\ d \simeq 18$$

∴ Population will be equal in 1972 + 18 = 1990

In that year the populations of A and B are both estimated to be 188.8 − 2.5 × 18‡

$$= 188.8 - 45 \simeq 144\,000$$

NOTES ON ANSWER

* Mid-point of series = mid-point of 1972.

† Note that the minus sign indicates a *downward* trend.

‡ Since the populations are equal, substitution in only one equation is needed—preferably the simpler equation.

2. (*a*) (*i*) *Seasonal variations*

Year	*Quarter*	*Demand (tonnes)*	*M.A.T.*	*Moving average*	*Centred average**	*Seasonal variation*
1974	1	218				
	2	325				
			1064	266		
	3	273			294	93
			1290	322½		
	4	248			355	70
			1550	387½		
1975	1	444			409	108½
			1722	430½		
	2	585			448	131
			1859	464¾		
	3	445			492	90½
			2075	518¾		
	4	385			552	70
			2342	585½		
1976	1	660			608	109
			2520	630		
	2	852			648	131½
			2660	665		
	3	623				
	4	525				

Quarter	*1*	*2*	*3*	*4*	*Total*
1974	—	—	93	70	
1975	108½	131	90½	70	
1976	109	131½	—	—	
Total	217½	262½	183½	140	
Average	108¾	131¼	91¾	70	401¾†
Adjusted	108	131	91	70	400

The seasonal variations are, therefore:

Quarter 1:	108	Quarter 3:	91
2:	131	4:	70

(*ii*) *The trend* ‡

$$y = a + bd$$

Year	*Demand* (y§)	d	yd	d^2
1974	1064	−1	−1064	1
1975	1859	0	0	0
1976	2660	+1	+2660	1
	$\Sigma y = 5583$		$\Sigma yd = 1596$	$\Sigma d^2 = 2$

$a = \bar{y} = 5583/3 = \underline{\underline{1861}}$

and $b = \Sigma yd/\Sigma d^2 = 1596/2 = 798$

$\therefore\ y = \underline{\underline{1861 + 798d}}$

(*b*) (*i*) *Total demand for 1978:*

1978 is 3 years' deviation from 1975

$\therefore$ demand $y = 1861 + 798 \times 3$

$= 4255$, (say) $\underline{\underline{\text{4250 tonnes**}}}$

(*ii*) *Demand for last quarter* (4) *of 1978:*

First adjust the trend formula for quarters, i.e.:

$y' = 1861/4 + 798/4^2 \times d'$ (where d' = Deviation in quarters)

$\therefore\ y' = 465 + 50d'$ (rounded).

Mid-point of series = Mid-point of 1975, i.e. end of Quarter 2, 1975.

$\therefore$ Deviation of Quarter 4, 1978, from the mid-point of the series:

	Quarters
From end of Quarter 2, 1975, to the end of 1975	2
Quarters of the years 1976 and 1977	8
From beginning of 1978 to end of Quarter 3 ..	3
End of Quarter 3 to mid-point of Quarter 4 ..	$\frac{1}{2}$
	$\underline{\underline{+13\frac{1}{2}}}$

$\therefore$ The trend for Quarter 4, 1978:

$y' = 465 + 50 \times 13\frac{1}{2} = \underline{\underline{1140}}$

$\therefore$ Demand for last quarter of 1978:

= Trend × Seasonal variation

$= 1140 \times 70 = 798$, (say) $\underline{\underline{\text{800 tonnes}}}$

NOTES: * These figures are rounded.

† A total adjustment of $-1\frac{3}{4}$ is needed. If this is made using formula in XXVI, **4** and the variations rounded, the figures on the next line are obtained.

‡ In practice, 3 years is too short a duration to obtain a reliable trend equation.

§ These figures are the sum of the four quarters for each year.

** Since the estimated figures cannot be accurate to a tonne, it is better to round to a reasonable number.

4. Regression line y on x formulae:

(a) $\Sigma y = an + b\Sigma x$

(b) $\Sigma xy = a\Sigma x + b\Sigma x^2$

Now x in a time series is the time variable and in such a series can be measured by d.

$\therefore\ \Sigma x = \Sigma d$

But since the deviations d are measured from the mid-point of the series, and the deviations are in equal steps, then Σd must be zero.

$\therefore\ \Sigma x = 0$

$\therefore$ The above regression line formulae become:

(a) $\Sigma y = an + b \times 0 = an$

$$\therefore\ a = \frac{\Sigma y}{n} = \underline{\underline{\bar{y}}}$$

(b) $\Sigma xy = a\Sigma x + b\Sigma x^2$

Substituting d for x:

$$\begin{aligned} \Sigma dy &= a\Sigma d + b\Sigma d^2 \\ &= a \times 0 + b\Sigma d^2 \\ &= b\Sigma d^2 \end{aligned}$$

$$\therefore\ b = \underline{\underline{\Sigma dy/\Sigma d^2}}$$

But these are the formulae used to compute a trend line.

$\therefore$ The trend line is the regression line of y on x.

APPENDIX VII

Examination Questions

1. "Britain has 86 000 alcoholics—but America's hair-raising figure is 4 589 000." Comment. *B.S.I.*

2. Membership of trade unions in the U.K., 1938, 1951 and 1960.

Membership	Number of trade unions		
	1938	1951	1960
Under 100	309	145	132
100–	286	191	176
500–	99	78	51
1 000–	116	101	97
2 500–	74	77	67
5 000	59	43	34
10 000–	21	26	18
15 000	21	19	26
25 000 and over	39	50	49
Total	1024	730	650

Write a report on the changes in the sizes of trade unions over the period, using the statistics given in the above table and any derived statistics you find useful for the purpose.

U.L.C.I.

3. Write a short statistical commentary on industrial stoppages in the transport and communications industry in the period 1952 to 1961, using the figures in the following table, and compute such derived statistics as you require. What further statistical information would you require in assessing whether the transport and communications industry lost more days than other industries through stoppages in this period?

INDUSTRIAL STOPPAGES IN THE UNITED KINGDOM

	Transport and communications			All industries and services		
	Number of stoppages	*Workers involved* (000s)	*Working days lost* (000s)	*Number of stoppages*	*Workers involved* (000s)	*Working days lost* (000s)
1952	55	11	32	1714	415	1792
1953	73	38	69	1746	1370	2184
1954	125	113	919	1989	418	2457
1955	118	153	1687	2419	659	3781
1956	102	20	35	2648	507	2083
1957	121	165	998	2859	1356	8412
1958	83	100	2116	2629	523	3462
1959	88	30	95	2093	645	5270
1960	179	153	636	2832	817	3024
1961	138	54	230	2686	771	3046

I.O.T.

4. Analyse the data given below by means of secondary statistics, calculating any percentages that you think necessary. Summarise your results in a brief report on the fire and accident insurance business.

THE "L" GENERAL ASSURANCE SOCIETY, LTD.
Fire and accident business
1962

	Fire	*Accident*
Premium income, 1962	£3,964,120	£4,043,551
Increase on 1961	8%	5%
Fund at end of 1962	£2,050,798	£2,971,547
Claims as percentage of premium	47%	56%
Expenses as percentage of premium	48%	35%
Underwriting surplus	£64,255	£249,318

L.C.C.

5. Explain the following statistical terms:

(*a*) Absolute error.
(*b*) Relative error.
(*c*) Biased error.
(*d*) What is the absolute error in the following amounts, which are quoted to the nearest £10 000:
£50 000 + £280 000 + £1 560 000 − £710 000?

(*e*) What is the relative error in the following amounts which are quoted to the nearest 10: 500 × 200. *A.C.C.A.*

6 (*a*) Define absolute and relative errors of rounding.

(*b*) Explain the addition, substraction, multiplication and division laws of errors.

(*c*) The table below gives sales turnover and number of articles sold by four different agents. The figures have been rounded to the nearest £1 000 for the sales turnover and the nearest 10 for the number of articles sold.

Agent	Number sold	Sales turnover
A	860	51 000
B	640	42 000
C	1,500	86 000
D	480	34 000

Calculate the average selling price for each agent and the group as a whole.

(*d*) Calculate the absolute error due to rounding in the figures for:

(i) Sales turnover,
(ii) Number sold,

for each of the four agents and the group as a whole.

(*e*) Calculate the relative error due to rounding in the price per article for the group as a whole.

7. Distinguish between "primary data" and "secondary data," giving *two* examples of each.

Explain why secondary data must be used with great care. *Inst.M.*

8. What is "quota sampling"? Explain its advantages and disadvantages compared with other sampling methods for survey work. *I.H.A.*

9. Discuss the advantages and disadvantages of the postal questionnaire as a method of collecting data. *L.G.E.B.*

10. Examine the importance of non-response in social surveys. *University of London B.Sc.*

11. Draft a brief questionnaire (no more than 10 questions), to be completed by a sample of out-patients, which will provide

information regarding the views of these patients on the organisation of the out-patient department.

NOTE: No questions regarding medical treatment as such are required. *I.H.A.*

12. A manufacturer of gas appliances wishes to carry out a survey of households in a given town in order to assess the success, or otherwise, of their recent conversion to natural gas.

You have been asked to advise on the following points:

(*a*) How big a sample should he pick?
(*b*) What should he use as a sampling frame?
(*c*) Should he use postal questionnaires or carry out interviews?
(*d*) Should he use open-ended or precoded questions?

Answer each of the questions briefly, stating in each case the considerations that have led to your decision. *I.C.M.A.*

13. Tabulate the following information which has been taken from the *Financial Times*, 1963.

There has been a substantial increase in exports of New Zealand meat to new and developing markets in the period 1st October 1962 to 31st May 1963. Shipments outside the U.K. totalled 86 255 tons or 27.7 per cent of the total shipments compared with 67 500 tons or 23.4 per cent shipped outside the U.K. in the corresponding period in 1962. Increased shipments were sent to almost all major markets including the U.K. Shipments to the U.K. totalled 225 000 tons (221 000), U.S. east coast 34 700 tons (20 900), west coast 14 500 tons (10 000), Canada east coast 6700 tons (2800), Japan 13 000 tons (5700). Meat export receipts through the banking system in the first 5 months of 1963 were almost £45 m., compared with nearly £37 m., at the same time the previous year.

Calculate the relevant ratios necessary to bring out the relationships present. *R.S.A.*

14. Arrange the following data in concise tabular form:

"At 10th June 1963, 479 713 persons were registered as unemployed in Great Britain. Of these 340 570 were men, 106 272 were women, 19 322 were boys under 18 years, and the rest were girls under 18 years. The numbers temporarily stopped were 13 794 men, 432 boys, 4117 women and 668 girls, the remainder being classed as wholly unemployed. Of this group of wholly unemployed, 47 335 men had been unemployed for not more than 2 weeks, 38 231 for more than 2 but not more than 5 weeks, 27 370 for more than 5 but not more than 8 weeks, and 213 820 for more than 8 weeks. The corresponding figures for boys were 5234, 3849, 2728, 7079; for women 14 097, 15 690, 12 202, 60 166; and for girls 3543, 2746, 1951, 4641." (Source: *Ministry of Labour Gazette.*)

Briefly describe the main features of the figures displayed in your table. *U.L.C.I.*

15. The following facts about three London Boroughs are taken from the 1966 sample census of England and Wales:

> There were 47 740 households with one car and 9900 with two or more cars in Bromley, 27 460 with one car and 3550 with two or more cars in Greenwich, 14 500 with one car and 910 with two or more cars in Tower Hamlets. The number of cars in the three boroughs was 68 740, 34 940 and 16 370 respectively.

An anlaysis of garaging arrangements showed that 10 970 cars were left at the roadside at night in Bromley, 13 800 in Greenwich and 9760 in Tower Hamlets.

The total number of households in each borough were: Bromley 96 990, Greenwich 74 050 and Tower Hamlets 64 320 and the number of people in these households 289 760, 220 850 and 187 580 respectively. The area of each borough is 39 293, 11 724 and 4994 acres respectively.

Consider the foregoing and then:

(*a*) draw up tables to present a concise summary of the information,

(*b*) calculate the percentage of households in each borough without a car,

(*c*) calculate the percentage of cars in each borough left at the roadside at night,

(*d*) say how far the variation in (*c*) is related to population density.

I.O.T.

16. On 15th August 1960, the unemployment rate in Great Britain for males was 49 700 who had been without work for less than 2 weeks, 121 300 over 8 weeks and 53 200 over 2 weeks and less than 8 weeks. The respective figures for females were 23 500; 38 300 and 27 500. Comparable figures for the 11th July were for

males 44 900; 122 700 and 37 900 and for females 17 300; 40 500 and 18 400. These figures include persons under 18 registered as unemployed who have had no insured employment: these numbered 205 000 and 202 300 for males in August and July respectively and for females 76 300 and 74 200 respectively.

Tabulate this information in a presentable form under a suitable title.

A.C.C.A.

17. Graph the following data in suitable form for a management committee. Draft a short statement drawing attention to the main features of the data.

	Average daily occupation of beds	*Whole time nursing staff*	*Cost of services £ millions*	*Prescriptions dispensed millions*	*Payments to pharmacists £ millions*
1951	52.2	18 521	32.1	20.8	5.3
1952	53.7	18 972	32.8	20.4	5.7
1953	54.2	19 625	33.9	20.5	5.2
1954	55.4	20 279	36.6	20.9	5.7
1955	55.4	20 738	40.1	21.2	5.8
1956	54.3	21 036	43.6	21.5	6.2
1957	54.6	21 586	45.6	20.9	7.1
1958	54.3	21 968	48.1	20.4	7.6

I.H.A.

18. Study carefully the figures given below, and illustrate what you consider to be their most interesting features by means of one or more suitable graphs accompanied by brief notes.

FISHING FLEETS AND LANDINGS OF FISH OF BRITISH TAKING

	1956	1959	1962	1965
Total fishing vessels:				
England and Wales	4 525	4 836	4 839	4 987
Scotland	3 304	3 136	3 063	2 937
Trawlers:				
England and Wales	1 875	2 049	2 098	2 102
Scotland	193	179	165	321
Creel fishing Scotland	1 075	1 050	1 199	1 216

Landings of fish of British taking:

(Th. tons landed weight)	1956	1959	1962	1965
All fish	948	875	807	902
Cod	387	336	315	311
Haddock	148	121	135	166
Herrings	133	133	88	97
(Value in £'000)				
All fish	48 011	52 228	51 107	60 696
Cod	18 621	20 719	19 177	25 024
Haddock	8 034	8 765	9 673	10 848
Herrings	2 845	2 837	2 731	2 898

Source: *Ministry of Agriculture, Fisheries and Food, Department of Agriculture and Fisheries for Scotland*

I.C.M.A.

19.

INDEX OF INDUSTRIAL PRODUCTION

Average 1963 = 100

Total all industries

1963	100.0
1964	108.4
1965	111.8
1966	113.2
1967	113.7
1968	119.8

Source: *C.S.O.*

(*a*) Plot the Index of Industrial Production on a ratio (semi-logarithmic) graph. On the same graph plot 3 lines, showing compound rates of increase over 1963 of 3 per cent, 4 per cent and 5 per cent per annum respectively.

(*b*) On the basis of the figures alone, does a forecast rate of increase in industrial production of 5 per cent per annum for the next five years seem realistic?

(*c*) Why is it easier to use a ratio graph rather than a natural graph for this type of analysis? *I.C.M.A.*

20. Describe the differences between a natural scale graph and a semi-logarithmic graph. Indicate the type of information which is better shown on a semi-logarithmic graph. *A.C.C.A.*

21. The following figures were taken from the *Monthly Digest of Statistics*:

BRITISH RAILWAYS

Year	*Total number of passenger journeys (millions)*
1950	981.7
1951	1001.3
1952	989.0
1953	985.3
1954	1020.1
1955	993.9
1956	1028.5
1957	1101.2
1958	1089.8
1959	1068.8
1960	1036.7
1961	1025.0
1962	965.0
1963	938.0
1964	927.6
1965	865.1

Source: *Ministry of Transport*

(*a*) Round the figures of the number of passenger journeys to the nearest 10 million.

(*b*) Calculate a 5-year moving average of the rounded series.

(*c*) Plot the figures on a graph.

(*d*) What is the maximum possible error due to rounding on each moving average? *I.C.M.A.*

22. What is a Lorenz curve? Write a brief explanation and use this technique to compare the two sets of figures tabulated below:

DISTRIBUTION OF PERSONAL INCOMES AFTER TAX
YEARS ENDED 5TH APRIL

Range of incomes £	1938/9 *Number of incomes (thousand)*	1938/9 £ *million*	1949/50 *Number of incomes (thousand)*	1949/50 £ *million*
125–149	2508	345	795	114
150–249	4630	871	6350	1274
250–499	1940	645	9897	3430
500–749	375	225	2040	1199
750–999	132	115	485	412
1000–1999	142	195	394	528
2000–3999	55	149	84	220
4000–5999	12	56	5	21
6000+	6	62	0	1
All incomes	9800	2663	20050	7199

Source: *Board of Inland Revenue*

What is your interpretation of the two curves? *I.C.M.A.*

23. Describe how a Lorenz curve is compiled.

Construct a suitable table of personal incomes and personal tax from which a Lorenz curve can be compiled.

NOTE: The figures shown in the table need not be accurate, but must clearly indicate the type of information found in such tables. *A.C.C.A.*

24. Describe the type of chart most suitable for illustrating the following:

(*a*) Sales in four areas of two products over the last four years;
(*b*) Prices of three materials over the last five years;
(*c*) The proportion of the total expenditure, during the previous year, of a bus company according to wages, fuel, maintenance, tyres, depreciation and other operating expenses;
(*d*) The traffic receipts of the British Transport Commission showing totals for British Railways, London Transport and Inland Waterways. *N.C.T.E.C.*

25. Draw charts to bring out the main features shown by the following figures of passenger movement. Say *briefly* what these main features are.

DESTINATION OF PASSENGERS LEAVING THE U.K.

Thousands

	1961	1962	1963	1964	1965
By sea:					
Europe and Mediterranean countries	2486	2631	2915	3024	3295
North America	121	128	127	115	108
Rest of world	159	169	169	169	154
By air:					
Europe and Mediterranean countries	2477	2716	3030	3400	3862
North America	398	459	528	670	782
Rest of world	164	183	210	260	322

Source: *Annual Abstract of Statistics*

I.O.T.

26. In September, 1968, the Department of Employment and Productivity conducted a survey into the distribution of gross earnings. The following results relate to the industry in which XYZ Limited, your company, operates:

FULL TIME MANUAL MEN

	Gross hourly earnings (shillings per hour)
Lowest decile	7.5
Lowest quartile	8.7
Median	10.6
Upper quartile	12.6
Highest decile	14.7
	(as percentage of median)
Lowest decile	71.2
Lowest quartile	82.6
Upper quartile	119.5
Highest decile	139.4

Use the figures below, which are abstracted from your company's payroll for the same period, to calculate comparable statistics for your own company.

What conclusions do you draw from the comparison?

GROSS HOURLY EARNINGS OF FULL TIME MANUAL MEN
XYZ LIMITED

Shillings per hour	*Number of men*
less than 5	20
5 and less than 7	490
7 and less than 9	970
9 and less than 11	2140
11 and less than 13	1580
13 and less than 15	1320
15 and less than 17	950
17 and less than 19	420
19 and over	110
Total	8000

I.C.M.A.

27. Estimate, by drawing a graph, the median and the lower quartile of each of the following distributions. Explain in your own words what these statistical measures represent and give brief comments on the changes between 1961 and 1964 as revealed by your graph and your estimates.

NUMBER OF MOTOR CYCLE, SCOOTER AND MOPED RIDERS
KILLED OR SERIOUSLY INJURED IN GREAT BRITAIN

Age Group	1961	1964
16 years and under	1649	2823
17 years	2208	3419
18–19 years	4011	4688
20–24 years	5586	4486
25–29 years	2332	1627
30 years and over	6718	5750

Source: Road Accidents, 1964

I.O.T.

28. In the week following the one described in Question 44, the record of deliveries of sand from the quarry is as follows:

Wt of load (*cwt*)	Under 10	10 and under 20	20 and under 30	30 and under 40	40 and under 50	60 and under 80	Total
No. of loads	50	61	121	78	18	2	330

Using the same axes and scales, draw two ogives to illustrate the deliveries of sand in these 2 weeks, distinguishing clearly between them.

Use your graph to find the median weight and the semi-interquartile range for each of the 2 weeks.

Compare the answer for the 2 weeks. *U.L.C.I.*

29. The weekly production figures of refrigerators of a manufacturing company over a period of 60 weeks are given below:

484, 499, 503, 465, 475, 478, 511, 525, 530, 504, 496, 509, 559, 543, 523, 517, 506, 565, 480, 491, 497, 513, 519, 525, 538, 528, 520, 575, 511, 530, 522, 549, 488, 500, 539, 523, 527, 520, 495, 455, 568, 545, 514, 518, 561, 475, 493, 472, 482, 521, 533, 460, 496, 534, 505, 585, 505, 501, 522, 527.

(a) Prepare a frequency table using appropriate class intervals
(b) Draw the histogram
(c) Calculate the mean weekly production
(d) Calculate the median and mode
(e) Discuss the relative advantages and disadvantages of mean, median and mode as measures to summarise a set of data.

I.A.A.

30. Membership of trade unions in the U.K., 1938, 1948 and 1960.

	No. of trade unions		
Membership	1938	1948	1960
Under 100	309	134	132
100–	286	217	176
500–	99	74	51
1 000–	116	102	97
2 500–	74	76	67
5 000–	80	60	52
15 500–	21	19	26
25 000 and over	39	51	49
Total	1024	733	650

Source: Annual Abstract of Statistics

31. What are the principles underlying the construction of a histogram? Explain the relations between the histogram, the frequency polygon, and the frequency curve. Do you consider

that the frequency polygon has any particular advantages over the other two forms of representation? Illustrate with a histogram the distribution of males which is given below, and use a frequency polygon to illustrate the distribution of females.

STUDENTS ADMITTED TO COURSES FOR A FIRST DEGREE

Age on admission (years)	*Males*	*Females*
17½–	15	27
18–	67	73
18½–	67	56
19–	41	49
19½–	22	20
20–	11	8
21–	5	5
22–	12	2

L.G.E.B.

32. In a group of 1459 university undergraduates, 1269 graduated and the rest failed to graduate. Using the data below, compare the arithmetic mean age on admission of those who graduated and of those who failed.

Age on admission (years/months)	*Graduated*	*Failed to graduate*
17/4	15	0
17/8	34	5
18/0	113	5
18/4	332	40
18/8	362	55
19/0	289	15
19/4	83	45
19/8	20	20
20/0	21	5

You should assume an upper age limit of 21 years.

L.G.E.B.

33. Using the data of Question 32 construct the appropriate cumulative frequency curves to estimate the median age of all undergraduates, and the median age of those who graduated. Estimate also the 20th and the 80th percentiles for each distribution. Comment on your results. *L.G.E.B.*

34. From the following figures taken from a price index, state what is:

(*a*) the median,
(*b*) the arithmetic mean,
(*c*) the lower quartile,
(*d*) the upper quartile,
(*e*) the quartile deviation,
(*f*) the mean deviation.

Explain the difference between the mean and quartile deviation.

53	116	70	54	36
70	98	176	139	38
67	102	161	77	137
93	133	56	93	81
81	131	185	88	54
86	92	83	60	109
95	24	117	35	57
106	19	171	81	79
105	59	55	110	186

A.C.C.A.

Compare these distributions by estimating the median, the upper quartile and the highest decile (tenth) for each.

University of London B.Sc.

35. The following figures show the number of passengers carried on each of 70 journeys by an aircraft with a seating capacity of 100. Compile a grouped frequency distribution from which you can estimate the following:

(*a*) The average percentage capacity used;
(*b*) If 63 passengers is the smallest profitable load, the proportion of flights which were unprofitable.

11	12	27	57	90	72	51	76	81	71
78	76	61	25	48	67	80	78	66	32
52	26	70	53	27	67	88	67	23	96
59	74	87	61	57	24	60	63	51	52
18	43	76	99	76	64	87	12	89	38
28	87	79	90	58	29	51	45	29	84
37	82	30	76	58	33	81	55	68	91

I.O.T.

36. Using the data of Question 31 find the arithmetic mean age of admission for males and females separately. State precisely

any assumptions that you make. Show how the two means may be combined to give the overall mean age for all students admitted. *L.G.E.B*

37. The following table concerns rates of pay and earnings by British Railways staff.

(*a*) Calculate the combined average rates of pay and average earnings of men and women at March 1958 and at April 1962.

(*b*) Find the percentage increases between the two years revealed by your results.

(*c*) Discuss the meaning of the difference between the increase in pay and the increase in earnings.

	Number of staff (000s)	*Average rates of pay*	*Average weekly earnings*
Male adults			
March 1958	502.9	£9 2*s*.	£12 0s.
April 1962	435.7	11 1	15 4
Female adults			
March 1958	32.8	6 18	7 7
April 1962	29.6	9 0	9 15

I.O.T.

38 Define dispersion.

Explain briefly three different methods of measuring dispersion.

The average weekly salaries of 10 workers is as follows:

£25.33, £30.12, £26.25, £25.00, £35.10, £28.95, £33.00, £24.20, £31.80, £27.50.

Calculate the:
(*a*) mean;
(*b*) mean deviation;
(*c*) standard deviation.

I.A.A.

39. The following table gives the years of construction of United Kingdom merchant vessels of 500 gross tons and over:

Year of construction	*No. of vessels*
January 1927 to December 1931	389
„ 1932 to „ 1936	150
„ 1937 to „ 1941	343
„ 1942 to „ 1946	907
„ 1947 to „ 1951	618
„ 1952 to „ 1956	634

(*a*) Tabulate the age distribution at December 1956 of the merchant vessels, commencing with the last group, 1952 to 1956.

(*b*) Calculate the mean age and the standard deviation of the age distribution. *N.C.T.E.C.*

40. WEEKLY INCOME OF HOUSEHOLDS IN THE U.K., 1960

Income	Number of households	
	renting accommodation	*in owned dwellings*
Under £3	33	16
£3–	276	95
£6–	228	122
£10–	378	178
£14–	594	299
£20–	445	369
£30–	138	153
£50 and over	15	52
Total	2107	1284

From: "The Family Expenditure Survey," *Ministry of Labour Gazette*, December 1961.

Compare the two income distributions by computing means and standard deviations.

University of London B.Sc.

41. Calculate for the following distribution the arithmetic mean and standard deviation:

INDUSTRIAL STATUS IN GREAT BRITAIN 1951. ANALYSIS BY AGE OF MALE EMPLOYEES

Age	15–19	20–24	25–29	30–34	35–44	45–54	55–64	64 and over
Number in 000s	1	5	18	33	122	127	85	50

I.H.A.

42. For the distribution in Question 41, estimate the median and quartile deviation. Explain what advantage, if any, these measures have compared with the mean and standard deviation.

I.H.A.

43. (*a*) Define the following statistical measures:

(*i*) Mean and median.

(*ii*) Standard deviation and range.

(*b*) The following figures give the radii of plastic stoppers. Calculate the mean and standard deviation of the radii.

Radii inches (thousandths)	*No. of stoppers*
415–	12
416–	25
417–	45
418–	53
419–	67
420–	50
421–	35
422–	26
423–	12
424–425	5
	330

L.C.C.

44. The manager of a sand quarry has kept a record of the deliveries of sand from the quarry during one week. The record is as follows:

Wt. of load (*cwt*)	Under 10	10 and under 20	20 and under 30	30 and under 40	40 and under 60	60 and under 80	Total
No. of loads	78	85	116	82	13	6	380

Calculate:

(*i*) the arithmetic mean of the weights,

(*ii*) the standard deviation of the weights.

Find the greatest possible percentage error in your answer for the arithmetic mean. *U.L.C.I.*

45. (*a*) Explain the meaning of absolute and relative measures of dispersion, and compare their use.

(*b*) Estimate a measure of variation in the monthly production over the first half year, for each of the following two factories.

Output of Component Parts (*thousands*)

	Jan.	Feb.	Mar.	Apr.	May	June
Factory A	36	28	24	27	32	33
Factory B	73	52	51	58	63	59

L.C.C.

46. State briefly what you understand by "skewness." For the following distribution, calculate the mean, the standard deviation and the median:

Steel rings sold

Diameter (cm)	Number (gross)
14.00–14.49	90
14.50–14.99	180
15.00–15.49	200
15.50–15.99	250
16.00–16.49	120
16.50–16.99	80
17.00–17.49	60
17.50–17.99	20

Use the figures you have calculated to obtain a measure of the skewness of the distribution. *I.C.M.A.*

47. Represent the following information on a scatter chart and use the chart to estimate the fare you would expect to be charged for:

(*a*) a 2000 mile journey;
(*b*) a 4500 mile journey to the United States;
(*c*) a 7000 mile journey.

STANDARD SINGLE FARES AND DIRECT DISTANCES BY AIR FROM LONDON

Destination	*Miles*	£	*p.*	*Destination*	*Miles*	£	*p.*
Karachi	4269	207	0	Rome	885	45	4
Beirut	2275	111	0	Reykjavik	1179	39	9
Hanover	434	24	1	Sao Paulo	5990	268	19
Helsinki	1156	60	1	Kano	2759	146	0
Istanbul	1553	79	9	Houston	4858	173	14
Leningrad	1331	77	18	Brazzaville	3959	177	0
Luxembourg	310	15	10	Benghazi	1707	72	16
Los Angeles	5438	191	8	Chicago	3954	152	13
Khartoum	3062	139	0				
Boston (Mass)	3265	130	15				
Copenhagen	610	34	7				
Brussels	210	11	18				

I.O.T.

48. The figures below are the results of a series of observations made at a factory during the loading of gas cylinders on to lorries.

Time taken to load (minutes)	*Number of cylinders in load*
8	4
11	19
12	14
12	11
18	20
19	32
20	15
24	29
29	34
30	39

Express the relationship between the time taken to load and the number of cylinders in the load in the form of a regression equation. *I.C.M.A.*

49. Explain what you understand by "regression" and how it may be illustrated graphically. Show that there are always two regression lines between any pair of variables, and explain the relations between them. Use the data below to illustrate the regression of:

y = annual income at age 40
on x = age at termination of full-time education.

What would be the meaning, in this case, of the other line of regression?

x	y	x	y	x	y
14	500	15	600	18	2000
14	525	16	1250	19	710
14	750	16	850	20	1900
15	630	16	1500	22	3000
15	810	17	1380	23	1860

L.G.E.B.

50. A destructive test on a range of metal castings has shown a fairly close correlation between measurement x and strength y.

Sample	Measurement x	Strength y
a	0.6	10
b	0.9	13
c	1.8	18
d	2.3	21
e	1.7	24
f	2.8	26
g	2.4	27
h	2.8	28
i	3.3	31
j	3.1	34
k	3.8	36
l	4.0	40

Using these results, compile a formula for estimating strength y from measurement x and predict the strength of castings with the following measurements of x: 1.0; 2.2; 2.5. *I.C.M.A.*

51. The figures given below were taken from the *Monthly Digest of Statistics*:

Year	*Index of Industrial Production Chemicals and Allied Industries*
1958	100
1959	111
1960	123
1961	125
1962	129
1963	139
1964	152
1965	159
1966	165
1967	171

Source: D.E.P.

(*a*) Using the method of least squares construct a regression line to fit the above series. (Hint: Treat 1958 as year 1, 1959 as year 2, etc.).

(*b*) Use the line to predict the value of the index for 1970.

I.C.M.A.

52. Explain in your own words the meaning of the following terms:

(*a*) correlation;

(*b*) regression;

(*c*) standard error of the mean.

For what purposes are such statistical methods and measures used? *I.H.A.*

53 Define correlation coefficient.

The table gives the cinema admissions and TV licences per 1000 of population for 10 towns.

Town	1	2	3	4	5	6	7	8	9	10
Cinema admissions	12.8	10.9	13.7	12.4	8.6	13.2	10.0	11.8	10.9	8.4
TV Licences	45	242	98	190	208	105	265	53	205	131

Calculate the coefficient of correlation. *I.A.A.*

54. The following figures give the average weekly output of coal in the United Kingdom and the number of civil servants for twelve successive quarters in 1950–3:

Output of coal (100 000 *tons*)	*Civil servants* (000s)
40	689
41	685
36	679
44	676
46	675
39	680
45	686
43	688
42	684
38	678
47	673
48	668

(*a*) Calculate the coefficient of correlation between the two sets of figures.

(*b*) Would you agree that these figures demonstrate that a decrease in the number of civil servants results in an increase in coal production? Give reasons for your answer. *B.S.I.*

55. The intelligence quotient of 12 salesmen was measured using two different methods x and y. The results were as follows:

Salesmen	1	2	3	4	5	6
Method x	99	91	87	106	125	120
Method y	104	112	85	102	131	142

Salesmen	7	8	9	10	11	12
Method x	75	86	112	125	102	130
Method y	70	84	104	108	99	146

(*a*) Calculate the correlation coefficient between the two methods x and y.

(*b*) Draw the scatter diagram and compare the two results.

R.S.A.

56. The research department of your company has tested seven types of valve to see how efficiently they performed. The results, together with the prices of the valves, are given below:

Valve	*Price* £	*Ranking by performance*
A	132	2
B	98	6
C	117	1
D	89	5
E	145	3
F	100	4
G	85	7

Show by calculating a rank correlation coefficient whether in this instance price is a good guide to performance. *I.C.M.A.*

57. Ten types of paint have been subjected to artificial "weathering" tests in the laboratory. The results are listed in the table below, together with those obtained under natural conditions over a longer period of time.

	Ranking obtained:	
Type of paint	*In the laboratory*	*Under natural conditions*
A	2	1
B	4	2
C	1	3
D	6	4
E	3	5
F	5	6
G	8	7
H	10	8
I	7	9
J	9	10

Is the laboratory test a reliable guide as to how a given paint will behave under natural conditions? *I.C.M.A.*

58 (*a*) Four components in a machine function independently of each other; the probability of breakdown of each is 0.1. Calculate the probability that the machine will break down in consequence of a failure in any one of the four components.

(*b*) Calculate the probability of drawing an ace and a picture card out of a pack of 52 playing cards in successive draws. What is the probability of drawing an ace and a picture card from the same suit?

(*c*) In a hospital maternity ward 18 of the last 20 infants born have been female. Calculate the probability that (i) the next birth and (ii) the next two births, will be male. Explain the reasons underlying your answer to an expectant mother in the ward.

(*d*) A firm of accountants employs nine students working a 5 day week of which 3 days are spent outside the office. Calculate the probability of finding all nine students in the office on the same week-day.

If these employees are organised into three groups consisting of three students working together, calculate the probability that all nine students will be in the office together.

A.C.A.

59 The probability that machine A will be performing a useful function in 5 years' time is $\frac{1}{4}$ while the probability that machine B will still be operating use fully at the end of the same periode is $\frac{1}{3}$.

Find the probability that in five years' time:

(*a*) both machines will be performing a useful function;
(*b*) neither will be operating;
(*c*) only machine B will be operating;
(*d*) at least one of the machines will be operating.

I.C.M.A.

60. An experiment has three possible but mutually exclusive results:—A, B, and C. The probability that result A, or B or C will take place has been computed.

Which of the following assignments of probability are acceptable:

(i) $P(A) = 0.34$, $P(B) = 0.29$, $P(C) = 0.47$
(ii) $P(A) = 0.17$, $P(B) = 0.59$, $P(C) = 0.24$
(iii) $P(A) = \frac{13}{80}$, $P(B) = \frac{1}{2}$, $P(C) = \frac{25}{80}$

A.C.A.

(*a*) Explain briefly the value of conditional probability calculations to the businessman.

(*b*) *Defective Electron Tubes per box of 100 units*

Firm	0	1	2	3 or more
Supplier A	500	200	200	100
Supplier B	320	160	80	40
Supplier C	600	100	50	50

From the data given in the above table, calculate the conditional probabilities for the following questions:

(i) If one box had been selected at random from this universe, what are the probabilities that the box would have come from Supplier A; from Supplier B; from Supplier C?

(ii) If a box had been selected at random, what is the probability that it would contain two defective tubes?

(iii) If a box had been selected at random, what is the probability that it would have no defectives and would have come from Supplier A?

(iv) Given that a box selected at random came from Supplier B, what is the probability that it contained one or two defective tubes?

(v) If a box came from Supplier A, what is the probability that the box would have two or less defectives?

(vi) It is known that a box selected at random has two defective tubes. What is the probability that it came from Supplier A; from Supplier B; from Supplier C?

A.C.A.

62. (*a*) The mean inside diameter of a sample of 400 washers produced by a machine is 8.92 millimetres and the standard deviation is 0.12 millimetres. Washers with inside diameters within the range of 8.74 to 9.10 millimetres are acceptable.

Calculate the percentage of defective washers produced by the machine, assuming the diameters are normally distributed.

(*b*) Five per cent of the units produced in a manufacturing process turn out to be defective.

Find the probability that in a sample of ten units chosen at random exactly two will be defective using:

(i) the binomial distribution,

(ii) the Poisson approximation to the binomial distribution.

I.C.M.A.

63. The average number of newspapers purchased by a population of urban households in 1972 was 400, and the standard deviation was 100. Assuming that newspaper purchases were normally distributed, what is the probability that households bought:

(*a*) between 250 and 500 papers

(*b*) less that 250 papers

(*c*) between 500 and 600 papers

(*d*) more than 500 papers?

A.C.A.

64. "The ultimate object of sampling is to generalise about the total population." Explain this statement. *I.H.A.*

65. The following table gives the length of life of 400 radio tubes.

Length of life (hours)	*Number of radio tubes*
300– 399	12
400– 499	32
500– 599	64
600– 699	76
700– 799	88
800– 899	60
900– 999	32
1000–1099	26
1100–1199	10
	400

Calculate:

(*a*) The average length of life of a radio tube;

(*b*) The standard deviation of the lengths;

(*c*) The percentage number of tubes whose length of life falls within the range mean $\pm$ 2 times the standard deviation.

N.C.T.E.C.

66. The following figures were abstracted from personnel records:

1968

Number of days absent in year	*Number of employees*
None	620
1 to 5	876
6 to 10	498
11 to 15	204
16 to 20	101
21 to 25	80
26 to 30	67
31 to 35	22
36 to 40	17
41 to 45	2
46 to 50	0
51 to 55	3
56 to 60	5
61 or more	5
	2500

(*a*) Draw a rough graph of the figures.

(*b*) Calculate the mean number of days absent per employee and the standard deviation.

(*c*) Is it valid to deduce that 68 per cent of employees were absent between *mean less standard deviation* and *mean plus standard deviation* days? If not, why not? *I.C.M.A.*

67. The computer division of a large company has been asked to quote a realistic turn-round time for a certain type of job. The time quoted must be such that it will only be exceeded in approximately one case in six.

An analysis of turn-round times for similar jobs in recent months gives the following distribution of times:

Turn-round time (hours)	*Number of jobs*
less than 2	20
2 and less than 4	39
4 and less than 6	61
6 and less than 8	41
8 and less than 10	25
10 and less than 12	14
12 and over	3

(*a*) Calculate the mean and standard deviation of the distribution.

(*b*) What turn-round should the division quote in order to comply with the above condition? *I.C.M.A.*

68. The following table gives the gross weekly income of a sample of households:

	1957	*1959*
Under £3	133	87
£3 „ „ £6	268	354
£6 „ „ £8	159	152
£8 „ „ £10	232	206
£10 „ „ £14	619	549
£14 „ „ £20	760	808
£20 „ „ £30	479	628
Over £30	186	308

Calculate the arithmetic mean and the standard deviation *for* 1957 *only*. Calculate the standard error of the mean and explain its significance on the assumption that the above data are a random sample from the population of all households. *I.H.A.*

69. In the course of an investigation aimed at reducing lead

times for a given part a record was made of the frequency of machine stoppages during one month. The results are given below:

Machine stoppages

Duration (minutes)	*Frequency by case:*		
	Breakdown	Adjustment/ maintenance	Waiting time
less than 5	—	12	31
5 and less than 10	—	16	24
10 and less than 15	3	23	8
15 and less than 20	4	10	4
20 and less than 25	12	8	5
25 and less than 30	8	7	3
30 or more	5	2	5
Total	32	78	80

Calculate the mean stoppage time due to each of the causes and give confidence limits for your estimates. Comment on your findings. *I.C.M.A.*

70. In a sample of 500 garages it was found that 170 sold tyres at prices below those recommended by the manufacturer.

(*a*) Estimate the percentage of all garages selling below list price.
(*b*) Calculate the 95 per cent confidence limits for this estimate and explain briefly what these mean.
(*c*) What size sample would have to be taken in order to estimate the percentage to within 2 per cent? *I.C.M.A.*

71. A manufacturer aims to make electricity bulbs with a mean working life of 1000 hours. He draws a sample of 20 from a batch and tests it. The mean life of the sample bulbs is 990 with a standard deviation of 22 hours. Is the batch up to standard? *B.S.I.*

72. A given type of aircraft develops minor trouble in 4 per cent of flights. Another type of aircraft on the same journey develops trouble in 19 out of 150 flights. Investigate the performance of the two types of machine and comment on any significant difference. *L.C.C.*

73. In 1965 the accounts section of your company issued 34 101 invoices, of which 3120 were subsequently found to contain errors. After a reorganisation of the department a series of samples of invoices were inspected.

The following figures show the errors found:

		Number inspected	*Number with errors*
1966	June	120	6
	September	115	7
	December	130	5

Has there been a significant improvement? *I.C.M.A.*

74. After corrosion tests, 42 of 536 metal components treated with Primer A and 91 of 759 components treated with Primer B showed signs of rusting. Test the hypothesis that Primer A is superior to Primer B as a rust inhibitor. *I.C.M.A.*

75. The following figures are derived from the replies of women respondents interviewed in a sample survey on personal savings.

Value of savings	*Age* 16–44	45 *& over*
	%	%
None	33	27
Under £50	20	13
£50–99	21	9
£100–199	8	5
£200–499	11	18
£500–999	4	10
£1000–5000	2	14
£5000 and over	1	4
	100	100
No. in sample	969	144

For the above distributions:

(*i*) Estimate the mean savings for each age group.

(*ii*) Indicate the main difficulties involved in estimating the means of such distributions.

(*iii*) On the assumption that both samples are random, test the hypothesis that there is no statistically significant difference between the average savings of the two age groups.

(*iv*) Explain briefly, in non-technical language, what is meant by a "statistically significant" difference between sample means. *C.A. (C.M.I.)*

76. In a sample of 540 wives of professional and salaried workers, 42 per cent had visited their doctor at least once during the preceding 3 months. During the same period of a sample of 270 wives of labourers and unskilled workers, 36 per cent had visited a doctor. By the use of an appropriate statistical test, consider the validity of the assertion that middle class wives are more likely to visit their doctors than the wives of working class husbands. *I.H.A.*

77. One group of 60 patients is treated with drug A, while another group of 80 patients is given drug B. After 14 days, 12 members of the A group were free of infection; 20 members of the B group were likewise free. The doctor in charge is convinced that treatment with drug B is superior. Comment on this view, explaining the basis of any statistical test you may employ. *I.H.A.*

78. In a sample of 569 wives of professional and salaried workers 45 per cent attended weekly the local welfare centre with their infants. For a sample of 245 wives of agricultural workers, the corresponding proportion was 35 per cent. Test the hypothesis that there is no difference between the two groups in respect of their attendance at such centres. Explain how you interpret your result. *I.H.A.*

79. The following data relate to measurements of components produced by two different machines. Is there any significant difference between the output of the two machines?

Machine	*Number measured*	*Average size*	*Standard deviation*
A	1200	31 in.	5 in.
B	1000	30 in.	4 in.

I.C.M.A.

80. What do you understand by the term "statistically significant"? Where needed illustrate your answer by hypothetical data. *I.H.A.*

81. (*a*) What do you understand by the t-distribution?

(*b*) From a random sample of 9 machine components an average life of fifteen months is expected with a standard deviation of two months.

(i) Calculate the limits between which the actual average life of the components could vary, based on this sample, at 95% and 99% confidence limits.

(ii) Considering that the components are expected to be of high quality, what comments would you make on the results?

I.C.M.A.

82. The expected lifetime of electric light bulbs produced by a given process was 1,500 hours. To test a new batch a sample of 10 showed a mean lifetime of 1,410 hours. The standard deviation is 90 hours.

Test the hypothesis that the mean lifetime of the electric light bulbs has not changed, using a level significance of: (*a*) .05
(*b*) .01

I.C.M.A.

83. (*a*) What does the χ^2 test test?

(*b*) The number of rejects in six batches of equal size were:

Batch	Number of rejects
A	270
B	308
C	290
D	312
E	300
F	320

Test the hypothesis that the difference between them is due to chance using a level of significance of 0.05.

I.C.M.A.

84. Bricks made in two kilns have been graded as facings (high quality), seconds (medium quality) and commons (rather poor quality).

The production of bricks in a particular period was as follows:

	Facings 000's	Seconds 000's	Commons 000's	Total 000's
Kiln A	24	43	13	80
Kiln B	31	57	32	120

It is the policy of the business to produce bricks of the highest quality possible and the number of seconds and commons produced is due to the efficiency or inefficiency of the process. Use the χ^2 test to determine whether the manager of kiln A is justified in claiming that he produces bricks of a higher quality than his colleague at kiln B. *I.C.M.A.*

85. Given the following data, use the chi-squared test to determine whether the number of accidents in a group of factories is independent of the age of the worker.

Number of accidents	*Age*		
	Under 21	21–44	45–65
0	120	360	220
1	40	28	2
2	13	5	2
3 or more	7	2	1

Explain the principle underlying the test and how you interpret the results.

86. In order to test the effectiveness of a drying agent in paint, the following experiment was carried out. Each of six samples of material was cut into two halves. One half of each was covered with paint containing the agent and the other half with paint without the agent. Then all twelve halves were left to dry. The time taken to dry was as follows:

Drying time (hours)

	sample number					
	1	2	3	4	5	6
Paint with the agent	3.4	3.8	4.2	4.1	3.5	4.7
Paint without the agent	3.6	3.8	4.3	4.3	3.6	4.7

Required:

Carry out a 't'-test to determine whether the drying agent is effective, giving your reasons for choosing a one-tailed or two-tailed test. Carefully explain your conclusions.

A.C.A.

87. The number of breakdowns that have occurred during the last 100 shifts is as follows:

Number of breakdowns per shift	0	1	2	3	4	5
Frequency:						
Expected number of shifts	14	27	27	18	9	5
Actual number of shifts	10	23	25	22	10	10

Show whether the manager is justified in his claim that the difference between the number of actual and expected breakdowns is due to chance. It has been customary to use a significance level of 0.05.

I.C.M.A.

88. The following figures were obtained from a series of quality control checks on engineering components, six items being measured at hourly intervals.

Time:	9 a.m.	10 a.m.	11 a.m.	12 noon	1 p.m.	2 p.m.
Sizes:	119.0	120.8	119.7	119.9	118.5	119.7
	120.9	119.8	120.1	119.2	119.6	118.9
	119.0	118.9	120.9	121.1	120.2	123.1
	121.3	119.2	118.0	120.3	122.3	121.2
	120.0	120.5	118.6	122.6	122.9	118.0
	121.0	120.2	120.3	119.3	120.1	119.7

Depict these data on mean and range charts, using the following limits:

For the mean chart: Inner limits 119.5 and 120.5
Outer limits 119.2 and 120.8

For the range chart: Inner limit 3.5
Outer limit 4.5

Comment on the results. *I.C.M.A.*

89. Batches of 100 components were taken at fixed intervals of time from a production line and tested. The following figures are the number of components found to be defective in each of the batches:

4, 3, 2, 0, 7, 5,
2, 4, 1, 3, 4, 6.

(*a*) Calculate the mean proportion defective.

(*b*) Draw up a quality control chart for samples of 200 of the components, such that the inner limit is equal to the mean plus one standard error and the outer limit is equal to the mean plus two standard errors.

(*c*) Could your chart be used for batches of 50 components? If not, why not?

I.C.M.A.

90. The table below shows the prices and the annual consumption of the major raw materials used in a particular brewery in 1958 and 1966:

	1958 Price per ton £	1958 Consumption tons	1966 Price per ton £	1966 Consumption tons
Malt	49	19 874	46	25 116
Hops	512	732	724	496
Sugar	46	1 865	51	2 486
Wheat flour	31	873	27	2 093

Calculate a current-weighted index number showing the overall change in raw material prices.

Why is it very unlikely that you would have obtained the same result if you had used a base-weighted index? *I.C.M.A.*

91. "An index number is simply an average and as such is subject to all the limitations of this particular statistic." Explain and discuss this statement with reference to any official index number known to you. *C.A.*

92. INDEX OF INDUSTRIAL PRODUCTION

	Weight	*Index* 1961 (*Average* 1958 = 100)
Total manufacturing	748	114.7
Mining and quarrying	72	93
Construction	126	118
Gas, electricity and water	54	116

(*a*) Explain the meaning of "Average 1958 = 100."

(*b*) What is the purpose of weights?

(*c*) Calculate the index of industrial production for 1961.

(*d*) If the index for gas, electricity and water subsequently rises to 138 and other data remain the same, find the change in the total index of industrial production. *U.L.C.I.*

93. The total advertising outlay by means of (1) the Press, and (2) television, are as follows:

	Press £ millions	*Television* £ millions
1958	158.4	48.0
1959	170.5	57.4
1960	197.2	71.8
1961	197.5	83.0
1962	202.0	82.5

(*a*) Take 1958 as base year and calculate indexes of the amounts spent on advertising in each of the given categories.

(*b*) Compare the two sets of indexes by means of a compound bar chart. *R.S.A.*

94. The second column of the following table lists the average quantities of three metals used by a certain factory in the period 1950–2. Calculate the weighted aggregate price index for these metals for 1951 and 1952 using 1950 as the base year:

1	2	3	4	5	6	7	8
	Typical quantities used	*Price per pound*			*Price of quantities used*		
		1950	1951	1952	1950	1951	1952
Metal	(w)	P_0	P_1	P_2	P_0w	P_1w	P_2w
Copper	20.000 lb	$0.242	$0.242	$0.242	$4 840	$4 840	$4 840
Lead	12.000 lb	0.170	0.190	0.148	2 040	2 280	1 776
Zinc	14.000 lb	0.175	0.195	0.125	2 450	2 730	1 750

B.S.I.

95. The following figures were published by the Ministry of Labour.

INDEX OF RETAIL PRICES

Food

January 1962 = 100

Year	*March*	*June*	*September*	*December*
1964	105.8	109.1	108.1	109.9
1965	110.4	112.5	111.7	113.3
1966	113.1	118.4	115.1	117.0
1967	117.5	121.8	116.7	120.1
1968	122.1	124.1		

(*a*) Calculate a four-quarterly moving average of the above series.

(*b*) Calculate seasonal corrections for each of the quarters.

(*c*) Apply the corrections to the above series to obtain a seasonally corrected series of food prices.

(*d*) Plot both the original series and the corrected series on the same graph and comment on your findings. *I.C.M.A.*

96. Describe concisely the steps involved in calculating, by the method of moving averages, a seasonally-adjusted series. Of what use is such a series in a transport undertaking? By analysing his receipts in recent years, a transport operator estimates that on average his takings each January are 20 per cent below trend, those in February are 10 per cent below, and those in March 5 per cent below. In January, February and March 1963, his actual takings were £1000, £1100 and £1150 respectively. After allowing for seasonal influences, would you conclude that his receipts were rising or falling during this period? *I.O.T.*

97.

HOUSING IN ENGLAND AND WALES

Number of permanent houses completed for:

		Local housing authorities	*Private owners*
1960	4	26 534	43 745
1961	1	22 153	38 443
	2	23 253	43 599
	3	21 459	43 673
	4	26 015	44 651
1962	1	24 416	36 201
	2	24 951	42 638
	3	25 875	43 534
	4	30 040	44 643
1963	1	13 737	27 363
	2	26 253	42 971

Source: *Ministry of Housing.*

(*a*) Rewrite the number of houses completed, to the nearest thousand.

(*b*) Calculate the trend from the approximated figures, by means of a four-quarterly moving average.

(*c*) Plot both series with their respective trends, on the same diagram. *R.S.A.*

98.

MANUFACTURERS' SALES OF GRAMOPHONE RECORDS

	£'00 000				
	1959	1960	1961	1962	1963
Jan.–Mar.	30	38	39	43	46
Apr.–June	28	27	30	33	38
July–Sept.	30	33	36	37	47
Oct.–Dec.	48	53	55	62	

Calculate values for:

(*a*) the trend of the sale of records, and

(*b*) the regular seasonal movement of the sales, using the method of moving averages. *Inst. M.*

99.

EARNINGS IN INDUSTRY (VALUES TO THE NEAREST SHILLING)

		Men	*Women*			*Men*	*Women*
1956	April	235	120	1960	April	282	145
	October	238	123		October	291	148
1957	April	242	126	1961	April	301	153
	October	252	130		October	307	155
1958	April	253	131	1962	April	313	157
	October	257	134		October	317	161
1959	April	263	137	1963	April	323	164
	October	271	141		October	335	168

(*a*) Plot the two series of earnings given above, using an arithmetic scale graph.

(*b*) Calculate the trend of earnings for men, using a moving average in two's and plot the trend on the same diagram as the original series. *N.C.T.E.C.*

100. Define in your own words what is meant by any *three* of the following statistical terms and discuss their uses. In each case, give an example of their use in the field of transport.

(*a*) Frequency distribution;
(*b*) Random sample;
(*c*) Index numbers;
(*d*) Moving average;
(*e*) Scatter chart. *I.O.T.*

Index